Introductory Fungal Pathogenicity

Introductory Fungal Pathogenicity

Ashwin Bhasi
P Thomas
Deepak Panda

2026

Agri Horti Press

New Delhi-110 084

ISBN: 978-93-49690-39-4 (Hardbound)

Published By

Agri Horti Press

New Delhi

E-mail: agrihortipress@gmail.com

Digitally Printed at : Replika Press Pvt. Ltd.

Preface

Fungi, as both microscopic organisms and visible agents of decay and disease, occupy a unique and complex position in the biological world. Their roles in ecosystems, agriculture, medicine, and industry are vast—and among them, the ability to cause disease in plants, animals, and humans is of particular scientific and practical importance. Understanding fungal pathogenicity is essential for managing crop loss, treating fungal infections, and exploring host-pathogen interactions at the molecular and ecological levels.

This book, titled Introductory Fungal Pathogenicity, is designed to offer a comprehensive yet accessible overview of the principles underlying fungal diseases and the mechanisms by which fungi interact with their hosts. It is intended for students, researchers, and professionals in plant pathology, microbiology, biotechnology, and related disciplines who seek a foundational understanding of fungal pathogenesis.

The book begins by introducing the biology of fungi, outlining their classification, morphology, and life cycles. It then delves into the various strategies fungi use to infect hosts, including adhesion, penetration, enzyme secretion, and immune evasion. Key concepts such as virulence factors, pathogenicity genes, host defense mechanisms, and the co-evolution of fungal pathogens and their hosts are discussed with clarity and scientific depth.

Subsequent chapters explore the diversity of fungal diseases across kingdoms—highlighting plant pathogens like Fusarium and Magnaporthe, as well as medically important fungi such as Candida, Aspergillus, and Cryptococcus. Each example is contextualized with case studies and current research findings to illustrate real-world relevance and ongoing scientific challenges.

In addition, Introductory Fungal Pathogenicity addresses the tools and techniques used in studying fungal diseases, including microscopy, molecular diagnostics, genome sequencing, and functional genomics. Attention is also given to disease management approaches, such as fungicide use, biological control, and host resistance breeding, with discussion on the growing concerns surrounding antifungal resistance and environmental sustainability.

As fungal diseases become more prominent due to climate change, global trade, and emerging pathogens, the need for a solid understanding of fungal pathogenicity is greater than ever. This book aims to provide that foundation while encouraging further inquiry and research.

We hope Introductory Fungal Pathogenicity serves not only as an educational resource but also as an invitation to explore the fascinating and impactful world of fungal pathogens.

Ashwin Bhasi
P Thomas

Deepak Panda

Contents

Chapter-1

Biofilm Formation as a Pathogenicity Factor of Medically Important Fungi

1. Introduction

Since the seventeenth century, biofilms have been described in multiple systems. Most bacteria preferentially grow as biofilms, in all self-sustaining aquatic ecosystems, and these sessile bacterial cells differ deeply from their planktonic counterparts (cells in suspension) [1]. The definitions of biofilm have evolved over the years, in parallel to the advances of the biology area and research studies on the subject. The definition used today was proposed by Donlan and Costerton in 2002, and it describes a biofilm as a microbial community in which the cells are connected to a substrate, or to each other, embedded in a extracellular matrix of polymeric substances (produced by themselves) and exhibit an altered phenotype regarding the rate of growth and transcription of genes [2].

In fungi, the ability to colonize surfaces and to form biofilms was initially demonstrated for *Candida albicans* and *Saccharomyces cerevisiae*, in the 1990s and early 2000s [3, 4]. However, the growing awareness on the importance of fungal biofilms can be confirmed by the increased number of publications upon biofilm formation by other *Candida* species [5–7], as well as other yeasts that cause opportunistic infections and pneumonia in humans, such as *Malassezia pachydermatis* [8], *Rhodotorula* sp. [9], *Trichosporon asahii* [10], *Blastoschizomyces* [11], *Pneumo- cystis* spp. [12] and *Cryptococcus neoformans* [13]. Moreover, the ability to form biofilms has also been demonstrated in several filamentous fungi, including *Aspergillus fumigatus* [14] and *Fusarium* spp. [15], in fungi that cause endemic mycoses such as *Histoplasma capsulatum* [16], *Paracoccidioides brasiliensis* [17] and *Coccidioides immitis* [18] and in zygomycetes such as *Mucorales* [19].

Until now, several superficial reports about the ability of a wide range of fungal species to form biofilms *in vitro* and *in vivo* have popped up, demonstrating that this is possibly due to a favorable lifestyle organization used by most medically important fungi. Deeper knowledge about those biofilms is still a challenge. As they account for the first and second leading fungal infections on hospitals, *Candida* and *Aspergillus* biofilms are the most studied examples. Therefore, the next sections of this chapter highlight the most important features of *Candida* and *Aspergillus* biofilms, as they are known up to this date. An additional section will sum- marize recently published features of biofilms of other species of fungi.

2. *Candida* spp. biofilms

Candida spp. are often identified as the causative agent of candidemia, hospital pneumonia and urinary tract infections and, almost invariably, these infections are associated with the use of a medical device and biofilm formation on its surface [20]. The most commonly colonized medical device is the central venous catheter (CVC), used for administration of fluids, nutrients and medicines [21]. The infusion fluid or the catheter may be contaminated, but, more often, yeasts are introduced from the skin of the patient or the hands of health professionals [21]. Alternatively, these yeasts can migrate into the catheter from a pre-existing lesion. However, if *Candida* spp. that colonize the gastrointestinal tract as a commensal start to develop a pathogenic behavior, they are able to penetrate the intestinal mucosa, spread through the bloodstream and, then, circulating yeast may colonize the catheter endogenously. This could be a common dissemination mechanism in cancer patients because cancer chemotherapy leads to damage to the intestinal mucosa [22]. In non-neoplastic patients, infected catheters are the most important source of bloodstream infections followed by widespread invasive candidiasis. The catheter removal is recommended in patients with disseminated *Candida* spp. infection to facilitate disinfection of the blood and to improve prognosis [23–25].

Candidemia and other forms of invasive candidiasis (i.e., infection involving normally sterile sites) are the most prevalent invasive mycoses worldwide [20, 26] with mortality rates close to 40% [27, 28]. *Candida albicans* is the most commonly isolated species; however, in the past few decades, several surveillance studies reported an increased incidence of infections caused by *Candida* non-*albicans* species (CNA), like *C. glabrata*, *C. parapsilosis*, *C. tropicalis*, *C. krusei*, *C. guilliermondi* and *C. lusitaniae* [29, 30]. This epidemiological swift is of utmost importance because resistance to fluconazole and echinocandins (two of the three antifungal options in the clinical practice) has been shown to be more common in CNA species compared with *C. albicans* [31], especially due to some CNA species that are inherently resistant to antifungals, such as *C. krusei* to fluconazole [31], or have a greater propensity to develop antifungal resistance, such as *C. glabrata* [32, 33]. *Candida* spp., including *C. albicans* and the main CNA species related to candidemia, can colonize surfaces and develop biofilms, as demonstrated by several *in vitro* and *in vivo* studies [6, 34–38]. *C. albicans* is the third leading cause of catheter-related infec- tions, the second main cause of colonization-followed-by-infection [39, 40] and the mortality rate in patients with candidemia associated to catheter use is as high as 41% [41, 42].

Candida biofilm development can be didactically described in four sequential steps (Figure 1): (a) adherence—initial phase, in which the yeast in suspension and those circulating (planktonic cells) adhere to the surface, first 1–3 h; (b) intermediate phase, concerning the development of biofilm, 11–14 h; (c) maturation phase, in which the polymeric matrix completely soaks all layers of cells adhered to the surface in a three-dimensional structure, 20–48 h; (d) dispersion, in which the most superficial cells leave the biofilm and colonize areas surrounding the surface, after 24 h [43].

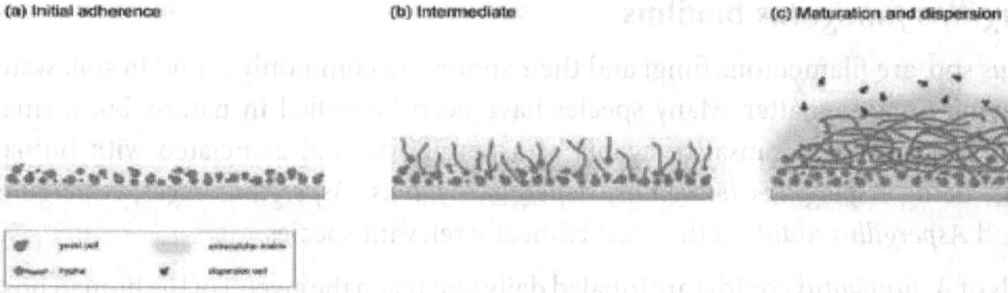

Figure 1. ***Candida*** **biofilm development model: (a) initial phase of adherence, in which the yeast in suspension (plank- tonic cells) adhere to the surface; (2) intermediate phase, accounts for growth of colonies and initial extracellular matrix secretion; (3) maturation phase, in which the ECM completely soaks all layers of cells adhered to the surface in a three- dimensional structure. After maturation, dispersion events, when the most superficial cells leave the biofilm and colo- nize areas surrounding the surface, may occur.**

The mature biofilm consists of a dense network of cells in the form of yeasts, hyphae and pseudohyphae (Figure 2A) soaked by polymeric extracellular matrix and with water channels between the cells, which facilitate the diffusion of nutrients from the environment through the biomass to the lower layers and which also allow the elimination of waste [43–45].

Biofilms of CNA species are less complex in structure because true-hyphae is not present, culminating in a biofilm formed predominantly by yeasts (*C. parapsilosis* and *C. glabrata* biofilms, Figure 2B and 2C, respectively) or, as observed for *C. tropicalis*, a mix of yeasts and some pseudo-hyphae [35].

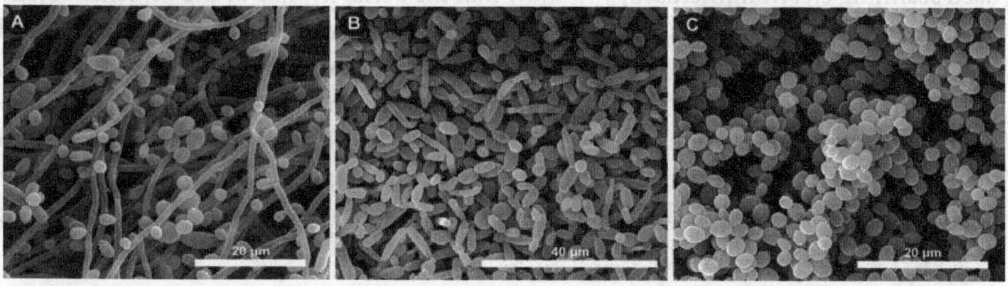

Figure 2. Scanning electron micrograph of ***in vitroCandida*** **spp. biofilms. (A)** ***Candida albicans*** **biofilms, formed by a dense network of cells in the form of yeasts, hyphae and pseudohyphae. (B)** ***Candida parapsilosis*** **biofilm, formed mainly by yeast cells, either round or elongated in shape; (C)** ***Candida glabrata*** **biofilm, formed exclusively by smaller, spherical yeast cells.**

Biofilms of *Candida* spp., formed using *in vivo* models, seem to follow the same sequence of *in vitro* formation [36]; however, maturation occurs more rapidly and the final thickness is increased. Mostly, *C. albicans* forms bi-layered biofilms, with a bottom layer formed of yeasts tightly attached to the surface and upper layers formed by hyphae; however, the final architecture of the biofilm is variable and depends, in part, on the substrate on which it is formed and on the growing conditions [43].

3. *Aspergillus fumigatus* biofilms

Aspergillus spp. are filamentous fungi and their spores are commonly found in soil, water and decaying organic matter. Many species have been identified in nature, but a small portion is recognized as causative agents of aspergillosis, and associated with human infections, being *Aspergillus fumigatus*, *Aspergillus flavus*, *Aspergillus niger*, *Aspergillus terreus* and *Aspergillus nidulans* the most clinically relevant species.

Hundreds of *A. fumigatus* conidia are inhaled daily and reach the alveoli of the human host. In immunocompetent individuals, conidia are efficiently eliminated by pulmonary macrophages. However, depending on the immune status of the host or predisposing conditions, *A. fumigatus* can lead to the development of disease in immunocompetent patients as in aspergi- loma patients with pre-existing pulmonary cavities or chronic obstructed sinuses (generated by tuberculosis, bronchiectasis or cystic fibrosis), in allergic rhinitis mediated by immunoglo- bulin E, in pneumonia and in allergic bronchopulmonary aspergillosis (ABA; clinical condition developed by patients with cystic fibrosis and asthma caused by *A. fumigatus* antigens). In immunocompromised patients, the pulmonary infection can spread into the bloodstream (invasive pulmonary aspergillosis; IPA) leading to the involvement of multiple organs. Invasive aspergillosis (IA) is the major infectious cause of morbidity in deeply immunocom- promised patients, especially post-transplant and/or with prolonged neutropenia; and mortality rates range from 40–90% [46]. *A. fumigatus* is responsible for approximately 90% of cases of IA [17, 47] and is, therefore, the most studied species. The initial establishment of chronic *A. fumigatus* infection involves the germination of conidia into mycelia and then subsequent invasion of the mycelial structure into pulmonary epithelial and endothelial cells [48]. In 2007, Beauvais et al. used scanning electron microscopy to show that the aerial hyphae of the mycelial colony formed over pulmonary cells were bounded together by a dense hydrophobic ECM, and that those colonies were more resistant to amphotericin B than liquid- submerged colonies, raising the hypothesis of biofilm formation during pulmonary *Aspergil- lus* colonization [49]. Confirmation was published by the same group, in 2010, using an *in vivo* model to demonstrate and characterize the presence of mature *A. fumigatus* biofilms (composed of hyphae covered with extracellular matrix) in aspergilomas and during the development of disseminated aspergilosis [46].

Following the first report in 2007, several studies demonstrated that *A. fumigatus* is able to grow as biofilms under *in vitro* conditions on polystyrene microtitre plates seeded with both human bronchial epithelial cells and cystic fibrosis (CF) human bronchial epithelial cells [50– 52]. Later on, *A. fumigatus* adherence and colonization of medical devices such as catheters, prostheses, cardiac pacemakers, heart valves and even breast implants have been extensively described [17, 53–55].

Compared to *C. albicans*, biofilm development is slower for *Aspergillus*, as a lag phase of approximately 10 h (conidial adhesion and germination) stands between the initial conidial seeding and the formation of an initial monolayer (early phase, 10–16 h). Then within the next few hours, intense hyphae grow and ECM secretion leads to increased structural complexity (intermediate, 48 h), culminating with a dense and mature biofilm after 72h (maturation phase) [50, 51].

Despite the lack of clinical studies substantiating *A. fumigatus* biofilm development *in vivo*, evidence such as high mortality in neutropenic cancer patients suffering from IA (40–90%) [46] and resistance of chronic infections to potent antifungal drugs *in vitro* [51, 56, 57] clearly indicates the formation of *A. fumigatus* biofilms *in vivo*. Additionally, histological and micro- scopic examination of bronchopulmonary lavage samples from the lungs have revealed the presence of numerous *A. fumigatus* hyphae in the form of dense intertwined mycelial balls or grains, referred to as mycetoma, which is similar to the biofilms formed by *Candida* species *in vivo* [36]. In fact, in 2009, Mowat et al. raised the discussion whether mycetomas should be considered biofilms [14].

4. Biofilms of other medically important fungi

Cryptococcosis, caused by yeasts of the genus *Cryptococcus* sp., is the third most prevalent disease in HIV-positive individuals. It is estimated that one million cases per year are associated with cryptococcosis in HIV-positive patients worldwide [58]. Infection by *Cryptococcus* occurs through inhalation of yeast spores in the environment and is considered a primary pulmonary infection that may progress to disseminated infection. Disseminated infection can affect the central nervous system (CNS), causing more severe forms of the disease like meningitis, encephalitis or meningoencephalitis [58]. More than 600,000 deaths are attributed to the 1 million new cases of cryptococcal meningitis that occur every year [59].

The main pathogenic species to humans are *C. neoformans* and *C. gatti*, with *C. neoformans* being the agent of opportunistic infections while *C. gatti* may also affect immunocompetent hosts. *Cryptococcus* sp. yeasts are able to colonize and form biofilms over various prosthetic devices such as peritoneal dialysis fistulas, hip prostheses and heart valves [55]. These biofilms include yeast cells with a vast amount of polysaccharide composing the extracellular matrix respon- sible for preventing its eradication by environmental agents and antimicrobials. Because *Cryptococcus* sp. in essentially an environmental fungus that adapted to the human host, biofilm formation is an expected survival strategy in harsh environmental conditions (e.g., ultraviolet light, dryness and natural antimicrobial substances). There are only a few studies on biofilms of *Cryptococcus* sp.; however, it is known that their formation is dependent on the presence of their polysaccharide capsule, mainly composed of glucuronoxylomanana (GXM), since anti-GXM antibodies specifically inhibit biofilm formation [60].

Invasive infections caused by *Candida* spp., *Cryptococcus* spp. and *Aspergillus* spp. are more frequently observed; however, other rare opportunistic fungi such as filamentous hyaline fungi (*Fusarium* spp., *Acremonium* spp. and species from the *Pseudallescheria/Scedosporium* complex) may also cause diseases that may vary from superficial to life-threatening invasive infections that may be fatal for immunocompromised individuals.

Fusarium species are common soil saprophytes and also important pathogens of plants and humans, causing superficial, invasive or disseminated infections. Twelve species are associ- ated with human infections, and *Fusarium solani*, *Fusarium oxysporum*, *Fusarium verticillioidis* and *Fusarium moniliforme* are the most important species in the human infection context [61]. As with aspergillosis, the clinical form of *Fusarium* depends on the immune status of the host. Among immunocompetent hosts, keratitis and onychomycosis are the most common infec- tions; therefore, in immunocompromised hosts, disseminated fusariosis is the second most common infection with filamentous fungus and affects especially patients undergoing therapy with high-dose corticosteroids with severe and prolonged neutropenia, in which a mortality rate up to 100% can be observed [61]. *Fusarium* spp. is also a major cause of microbial keratitis, and the formation of biofilms has been suggested as a contributing factor in recent outbreaks, especially associated with the use of contact lenses [62]. In addition to eye infections, *Fusari- um* sp. is also commonly isolated as the causative agent of onychomycosis. In nails, fungal cells generally form thick biomasses, containing embedded elements in a fungal extracellular matrix [63]. Several factors, including the firm adhesion to the nail plate, the presence of "persister cells" and the difficulty of eradicating the infection, suggest that biofilms are an important factor in the pathogenesis of onychomycosis [64]. *Fusarium* biofilm formation on polyestyrene surfaces, contact lenses and over human fingernails has been demonstrated *in vitro* and *in vivo* [15, 65–67] and may possibly occur on other medical devices, contributing to the high virulence and mortality observed in invasive infections.

5. Biofilm resistance behavior

Currently, antifungal therapy is based on four major classes of antifungal drugs: the polyene agents, azoles, allylamines and echinocandins. However, the therapeutic arsenal is limited by several problems, including selectivity, toxicity and development of resistance. Considering invasive mycoses, options are even more restricting, comprising amphotericin B, fluconazole (with several restrictions), itraconazole and voriconazole being the most suitable drugs. Although amphotericin B is considered to be the gold standard drug for these infections, its high degree of hepatotoxicity and nephrotoxicity [68] may turn it unacceptable for most patients predisposed to invasive fungal infections. Furthermore, some *Candida* species such as *C. krusei* and *C. glabrata* show less susceptibility to azole agents, which can lead to a therapeutic failure and often to death of the patient [32, 33]. From a clinical standpoint, resistance is the persistence or progression of an infection despite adequate medical therapy [69].

Fungal infections associated with biofilm formation are often poorly susceptible or even refractory to conventional antifungal therapies, which implies the need for higher dosages— not always possible, as discussed above—or antifungal combination therapy for better penetration of drugs in biofilms. The ineffectiveness of the azole antifungals and classical formulations of amphotericin B (deoxycholate) against biofilms of *Candida* spp. was demon- strated by several groups over the past few years [70–72], whereas only the echinocandins and lipid formulations of amphotericin B showed good activity against biofilms of *C. albicans* and *C. parapsilosis* [71]. Similarly, *Cryptococcus neoformans* and *C. laurenttii* biofilms were resistant to all tested azoles (itraconazole, fluconazole and voriconazole) [73, 74], but were susceptible to amphotericin B [74]. Importantly, biofilms of *A. fumigatus* were resistant to both voriconazole and echinocandins (anidulafungin and caspofungin) in two published studies [57, 75], being amphotericin B the only available antifungal drug with demonstrated activity against *A. fumigatus* biofilms available for clinical use [75]. Finally, *F. solani* and *F. oxysporum in vitro* biofilms also showed reduced susceptibility to all tested antifungal agents, including ampho- tericin B, voriconazole, itraconazole and fluconazole [66, 76]. Thus, the current scenario shows the scarcity of drugs available for the treatment of invasive fungal infections derived from biofilms, which are increasingly frequent in the hospital environment and frequently associ- ated with severe clinical conditions.

6. Biofilm mechanisms of resistance

According to the definition of a biofilm, the cells that compose this structure have an altered phenotype and differ from the planktonic cells (free-floating cells) in the expression of genes, rate of growth and also in its susceptibility to antifungal agents. The increased resistance to antifungals in *Candida* spp. grown as biofilms, in comparison to its planktonic forms, is the most medically relevant behavioral change associated to biofilms in the clinical setting [77]. Multiple mechanisms have been suggested to explain the increased antifungal resistance of the biofilm, including cell density, alteration of drug targets, expression of drug efflux pumps, the extracellular matrix and presence of persistent cells [55, 77–81]. Each of these mechanisms will be addressed separately in the next paragraphs, in the context of our chosen biofilm model (*C. albicans*), and recent finds concerning other fungi will be inserted when appropriate.

7. Cell density

The biofilm architecture is highly ordered to allow the infusion of nutrients and waste expulsion. Mature biofilms, even having high cell density, exhibit spatial heterogeneity with microcolonies and water channels, common feature of both biofilm bacteria and fungi [55]. It has been shown that both planktonic cells and cells resuspended from biofilms exhibit sensitivity to azoles when the cell density is low (10^3 cells/ml) and became more resistant when cell density is increased ten-fold [78]. It is believed, therefore, that the cell density is an important resistance factor within complex biofilms, particularly to azoles.

8. Drug target alteration

The antifungal agents of the azole class, including fluconazole, itraconazole, voriconazole and posaconazole, act by inhibiting sterol 14-α-demethylase enzyme encoded by ERG11 gene. The main target of azoles, Erg11p protein, can develop point mutations or be overexpressed, reducing the drug activity and culminating in an ineffective treatment. Treatment of *C. albicans* biofilms with fluconazole induces upregulation of genes encoding enzymes involved in the ergosterol biosynthesis (CaERG1, CaERG3, CaERG11 and CaERG25), this feature being even more pronounced in biofilms exposed for longer periods (22 h). Yet, treatment of *C. albicans* biofilms with amphotericin B results in increased expression of CaSKN1 predomi- nantly and a modest upregulation of CaKRE1 (both related to the cell wall) [82]. Upregulation of genes from the ergosterol biosynthetic pathway were also reported in biofilms of *C. dubliniensis* [83] and *C. parapsilosis* [84] and in a *in vivo* model of *C. albicans* biofilms using central venous catheters [85]. Additionally, the analysis of sterol composition of the biofilm cells of *C. albicans* has shown that the levels of ergosterol (the main sterol of fungal cell membrane) were significantly lower in the intermediate stages (12 h) and maturation (48 h) compared with the initial phase (6 h) of biofilm development [81]. Changing ergosterol exposition in the membranes of biofilm cells could explain their resistance both to azole agents as to polyenic, targeting the ergosterol molecule.

9. Drug efflux pumps expression

The primary molecular mechanism leading to resistance to the azoles, in *C. albicans*, is the increased efflux of the drug, mainly mediated by transporters from the ABC family and the MFS facilitators superfamily. The ABC transporters (ATP Linked), in *C. albicans*, constitute a multigene family, which includes multiple genes CDR (CDR1-4). Among the MFS family members, whom are secondary carriers and use the proton motive force, the MDR1 gene encodes an important mediator, which has been implicated in the resistance of *C. albicans* exclusively against fluconazole [55]. Various antifungal agents may be substrates for these pumps, and, therefore, its overexpression can lead to cross-resistance between different drugs, particularly azoles.

The increased expression of genes encoding drug efflux pumps has been reported in *C. albicans* [77, 79, 81], *C. glabrata* [86] and *C. tropicalis* [5] biofilms. Interestingly, the expression of CaCDR1, CaCDR2 and CaMDR1 is differentially regulated during development of the biofilm and after its exposure to antimicrobial drugs [77, 81, 87, 88]. Using *C. albicans* single, double and triple mutants for the main efflux pump genes (Δcdr1, Δcdr2, Δmdr, Δcdr1/Δcdr2 and Δmdr/Δcdr1/Δcdr2), Mukherjee and colleagues (2004) demonstrated that 6 h after formation, biofilms of double and triple mutants were 4–16 times more sensitive to fluconazole than biofilms of the wild type, while the biofilm from all strains become highly resistant to this azole after 12 and 48 h of development [81]. The lack of involvement of efflux pumps in mature biofilm resistance has been previously demonstrated by Ramage et al., also using *C. albicans* strains [77]. Collectively, the available literature supports the hypothesis that efflux pump overexpression is an important, but not exclusive, determinant of fungal resistance to azoles biofilms and may play an important role in the initial phases of biofilm development. Their primary function may be to allow the first cells to establish within complex environments and to protect them from acute toxicity, thus ensuring the permanence of these cells and allowing the biofilm to start to grow [81]. In the clinical setting, early exposure to azoles can, then, increase the expression of efflux pumps in early-established cells and contribute to induce clinical resistance.

10. Role of the extracellular matrix of the biofilm in resistance

In most biofilms, the population of microorganisms corresponds to 10% of the total mass and the extracellular matrix (ECM) corresponds to 90%. The ECM is a key biofilm component, which exerts a physical barrier function, protecting the cells from environmental factors such as host immunity and antifungal agents [21]. In 2004, Al-Fattani and Douglas demonstrated that, although the diffusion of small molecules can be hampered by the presence of a dense ECM, reducing the penetration of antifungal drugs does not play a key role in biofilm resistance [89]. Recent studies have provided new insights suggesting that the chemical composition of the ECM and its regulation may play the central role in resistance.

The overall composition of the ECM of *C. albicans* biofilms was first characterized by Baillie and Douglas [90] and confirmed later by Al-fattani et al. [89]. Recently, an extensive analysis of the ECM composition of *C. albicans* biofilms was published, where proteins appear as the major component (55%), followed by carbohydrates (25%), lipids (15%) and nucleic acids, mostly e-DNA (5%) [91]. Nuclear magnetic resonance (NMR) of exopolysaccharide fractions detected three major polysaccharides, similar to those found in the cell wall, but in quite different relative abundance. While β-1,3-glucan is the most abundant polysaccharide in the cell wall of *C. albicans* planktonic cells, the amount of β-1,3-glucan present in the ECM of its biofilms was surprisingly low. The most abundant polysaccharides are, actually, mannans and α-1,6-1,2-branching mannans, which appears to be associated with β-1,6-glucans, forming a glucan-mannan complex [91]. Much less is known about the ECM composition of biofilms of other *Candida* species and or fungi. Therefore, ECM may also resemble cell wall components in other species, as demonstrated for *A. fumigatus*, in which ECM is composed of galacto- mannan, α-1,3-glucans, galactosaminogalactan, monosaccharides and polyols, melanin and proteins [49].

The contribution of the β-1,3-glucan for the biofilm resistance in *C. albicans* was confirmed by a series of studies, which demonstrated that (*i*) the digestion of β-1,3-glucan residues by the addition of β-1,3-glucanase significantly improved the *in vitro* anti-biofilm activity of both fluconazole and amphotericin B drugs; (*ii*) the addition of exogenous ECM and/or β-1,3-glucan residues reduced the *in vitro* antifungal activity of fluconazole against *C. albicans* planktonic cells [92], resembling a biofilm-like behavior by the presence of the ECM. Also, β-1,3-glucan is responsible for sequestering all major drugs from the ECM environment, including azoles, echinocandins, polyenes and pyrimidines [93–95], behaving like a "drug sponge" and contributing to the increased resistance of the biofilm. A recent study published by Dr. Andes group (2014) upon this subject suggests that the most abundant polysaccharide in the ECM is not β-1,3-glucan (as previously thought), but a polysaccharide complex comprising an association of glucan-mannan residues, which is also capable of binding to fluconazole molecules and contributes to the resistance [91]. The work emphasizes that, most possibly, a large proportion of polysaccharides in the ECM may act as drug-sequestering molecules and contribute to biofilm resistance to antifungal agents.

In addition to polysaccharides, the extracellular DNA (eDNA) present in ECM of *C. albicans* biofilms also appears to have a role in resistance to non-azole agents. This feature was confirmed by Martins and colleagues, in 2010 and 2012, using DNase enzymes in association with antifungal drugs and confirming that destroying the eDNA with the enzyme led to an increased *in vitro* anti-biofilm activity of polyenes and echinocandins, but not azoles [96, 97]. Studies by Rajendran et al. have now also demonstrated that eDNA is also an important structural constituent of *A. fumigatus* ECM and plays an important functional role in main- taining the structural and architectural integrity of its biofilms. Furthermore, in this species, the release of eDNA by autolysis in biofilms is significantly associated with the levels of antifungal resistance, suggesting that eDNA plays an important role in *A. fumigatus* biofilm resistance to antifungals [98].

Other than physical components, transcription factors that regulate glucan synthesis and hydrolases are also associated with biofilm resistance. The CaZAP1 transcription factor is a negative regulator of the release of soluble β-1,3-glucan for the ECM in *C. albicans* biofilms. Yet, a group of alcohol dehydrogenases (CaADH5, CaCSH1 and CaLFD6) is associated with the production of ECM as they act as "quorum sensing" molecules, coordinating the matura- tion of biofilm [99]. In general, ECM production in *C. albicans* biofilms is highly regulated and is a key factor for resistance.

The ability to form *in vitro* biofilms containing ECM and its participation in the resistance has been described in other *Candida* species, including *C. glabrata*, *C. parapsilosis*, *C. tropicalis*, and *C. dubliniensis* [35] and, also, in other fungi, such as: *Cryptococcus neoformans*, *C. gattii*, *Pneu- mocystis* spp., *Blastoschizomyces capitatus*, *Malassezia pachydermatis*, *Saccharomyces cerevisiae*, *Rhizopus oryzae*, *Lichtheimia corymbifera*, *Rhizomucor pusillus* and *Apophysomyces elegant*, *Rhodotorula* spp., *Aspergillus fumigatus*, *Histoplasma capsulatum*, *Paracoccidioides brasiliensis*, *Coccidioides immitis*, *Fusarium*, *Trichosporon asahii* and *Mucorales* (Revised in [17, 55]), corrobo- rating the hypothesis that the ECM plays a critical role in fungal resistance and is one of the most significant mechanisms and regulated in the resistant phenotype of biofilm.

11. Persister cells

Persister cells are an important mechanism of tolerance in chronic infections and recently have received special attention in fungi biofilms [55]. By definition, these cells are "dormant variants of regular cells inside a microbial population that are highly tolerant to antibiotics" [100]. The main disruptive effect of antifungal agents in the cells relates to its interference with metabolic processes (synthesis of cell membrane, cell wall or DNA). The main characteristic of a "dor- mant" or "persister" cell is the reduction of its metabolism and cell division. So, because they are not metabolizing substrates and not dividing, these cells are no longer a target for the antifungal and become tolerant to its presence [100]. The presence of persister cells has been demonstrated in biofilms of *C. albicans*, *C. krusei* and *C. parapsilosis* treated with amphotericin B [101]. In an evidentiary study, re-inoculation of biofilm cells that survived the treatment with amphotericin B produced a new biofilm with a new subpopulation of persistent cells, sug- gesting that they were not mutants, but phenotypic variants of the wild type and adhesion on the substrate has triggered the formation of a persister subpopulation. Thus, in this clinically relevant scenario, inefficient and prolonged antifungal therapy may be beneficial for this subpopulation of the biofilm, which may be responsible for the ineffectiveness of the treatment and relapses [55].

In summary, the major studies published to date that attempt to elucidate the main factors involved in antifungal resistance of biofilms were performed with *C. albicans* biofilms and therefore, little is known about the specific resistance mechanisms for biofilms of other *Candida* species, or other biofilm-former fungi. It is likely that the ECM also acts as a barrier to the penetration of antifungal in those biofilms; however, as the ECM composition is different for each species, the role of the resistance to antifungal agents probably will not be the same. Likewise, the patterns of gene expression and sterol metabolism membrane will also be specific for each species.

12. Biofilm and pathogenesis

Pathogenesis involves the interaction between the pathogen and the host. To cause disease, the infectious agent makes use of both invasiveness factors—the pathogen virulence factors—and the ability to resist and evade the host immune system. Often these two topics commu- nicate, mainly because the molecules and metabolic adaptations produced by the pathogen to escape the immune response are considered as virulence factors.

The ability to grow as a biofilm cannot be considered a *classic* virulence factor, as the definition of virulence factor states that lack of the featured characteristic leads to non-virulent strains. Several fungi that do not form biofilm are still able to cause infection; however, those who do grow as biofilms are constantly linked to severe disease. Interestingly enough, a new molecule that impairs *C. albicans* biofilm formation does so by inhibiting the filamentation, an important virulence factor of this species. *In vivo* inhibition of filamentation and consequently biofilm formation depletes oral infection of immunocompromised mice [102]. This corroborates to the hypothesis that biofilm formation might be an important pathogenic factor and, thus, an important drug target.

The relationship between biofilm and pathogenicity relies mainly on two unique features of this community life-style: its increased resistance and the dispersion of infectious cells. Biofilms are a natural survival strategy of microorganisms to resist environmental threats [2]. In the clinical setting, the encased highly dense colony of fungal cells is protected not only from antifungal penetration, as discussed above, but also from the immune system. A single yeast or hyphae cell can be recognized and eliminated by the innate immunological response, either via phagocytosis by macrophages or induction of apoptosis by degranulation of mast cells. However, biofilms are too big to be phagocytosed and, yet, ECM may impair recognition of fungal surface epitopes. Thus, biofilm formation may also contribute to the escape from the host immunological response, favoring the establishment of the infection.

Candida mature biofilms (and possibly all fungi biofilms), after reaching a critical biomass, find a dynamic equilibrium in which the increase in cell density is offset by the release of superficial yeasts from the top, in a phenomenon called dispersion.

Cells that are released from mature biofilms are called "dispersion cells" and may colonize adjacent surfaces, expanding the biofilm or, in a clinically relevant scenario, use the blood- stream to disseminate the infection and allow the colonization of deep organs [103]. Addi- tionally, *C. albicans* dispersion cells exhibit significant phenotypic changes and are more virulent than those grown as planktonic (non-biofilm) cells. Alterations include: increased adherence to polystyrene, significantly higher germ tube formation, which is important because filamentation is essential for *C. albicans* virulence, more robust biofilm formation and increased virulence in a murine model of disseminated candidiasis [103]. Therefore, when a catheter is infected with fungal biofilm, "dispersion cells" with increased virulence potential may gain access to the bloodstream and disseminate the infection.

Recently, a prospective analysis of patients with *Candida* bloodstream infection (BSI) per- formed in Scotland confirmed that biofilm formation is a risk factor for mortality in patients with disseminated *C. albicans* infection [42]. Several previous works also showed that removal of a catheter within the first 24 h of candidemia diagnosis improves the clinical outcome and results in a shorter duration of candidemia with decreased mortality [42, 104, 105], confirming that biofilms function as reservoirs and are directly correlated to the dissemination of the infection. In fact, the current *guideline for the management of catheter-associated infections and their clinical management* states that, where possible, the catheter should be removed in non- neutropenic patients [23, 24, 106, 107].

13. Final considerations

The ability to form biofilms is widespread among pathogenic fungi, but understanding of the mechanisms that govern their formation, physiology and drug resistance is still limited. The continuous development of knowledge of the molecular mechanisms underlying biofilm formation, maintenance and molecular basis of metabolic dormancy of subpopulations of cells, such as persister cells, could lead to a drug-based strategy that could help us solve clinical diseases associated with fungal biofilms.

References

1. J. W. Costerton, Z. Lewandowski, D. E. Caldwell, D. R. Korber, and H. M. Lappin-scott, "Microbial biofilms," *Annu. Rev. Microbiol.*, vol. 49, pp. 711–745, 1995.
2. R. M. Donlan and J. W. Costerton, "Biofilms: survival mechanisms of clinically relevant microorganisms," *Clin. Microbiol. Rev.*, vol. 15, pp. 167–193, 2002.
3. S. P. Hawser and L. J. Douglas, "Biofilm formation by Candida species on the surface of catheter materials in vitro," *Infect. Immun.*, vol. 62, pp. 915–921, 1994.
4. T. B. Reynolds and G. R. Fink, "Bakers' yeast, a model for fungal biofilm formation.," *Science*, vol. 291, no. 5505, pp. 878–881, Feb. 2001.
5. F. C. Bizerra, C. V. Nakamura, C. De Poersch, T. I. Estivalet Svidzinski, R. M. Borsato Quesada, S. Goldenberg, M. A. Krieger, and S. F. Yamada-Ogatta, "Characteristics of biofilm formation by Candida tropicalis and antifungal resistance," *FEMS Yeast Res.*, vol. 8, no. 3, pp. 442–450, 2008.
6. A. A. Lattif, P. K. Mukherjee, J. Chandra, K. Swindell, S. R. Lockhart, D. J. Diekema, M. A. Pfaller, and M. A. Ghannoum, "Characterization of biofilms formed by Candida parapsilosis, C. metapsilosis, and C. orthopsilosis," *Int. Jl Med. Microbiol.*, vol. 300, no. 4, pp. 265–270, 2010.
7. S. Silva, M. Negri, M. Henriques, R. Oliveira, D. W. Williams, and J. Azeredo, "Adherence and biofilm formation of non-Candida albicans Candida species," *TRENDS Microbiol.*, vol. 19, no. 5, pp. 241-247, 2011.
8. F. T. Cannizzo, E. Eraso, P. a Ezkurra, M. Villar-Vidal, E. Bollo, G. Castellá, F. J. Cab- añes, V. Vidotto, and G. Quindós, "Biofilm development by clinical isolates of Malas- sezia pachydermatis.," *Med. Mycol.*, vol. 45, no. 4, pp. 357–61, Jun. 2007.
9. J. M. Nunes, F. C. Bizerra, R. C. E. Ferreira, and A. L. Colombo, "Molecular identification, antifungal susceptibility profile, and biofilm formation of clinical and environmental Rhodotorula species isolates.," *Antimicrob. Agents Chemother.*, vol. 57, no. 1, pp. 382–389, Jan. 2013.
10. G. Di Bonaventura, A. Pompilio, C. Picciani, M. Iezzi, D. D'Antonio, and R. Piccolomini, "Biofilm formation by the emerging fungal pathogen Trichosporon asahii: development, architecture, and antifungal resistance.," *Antimicrob. Agents Chemother.*, vol. 50, no. 10, pp. 3269–3276, Oct. 2006.
11. D. D'Antonio, G. Parruti, E. Pontieri, G. Di Bonaventura, L. Manzoli, R. Sferra, a Ve- tuschi, R. Piccolomini, F. Romano, and T. Staniscia, "Slime production by clinical iso- lates of Blastoschizomyces capitatus from patients with hematological malignancies and catheter-related fungemia.," *Eur. J. Clin. Microbiol. Infect. Dis.*, vol. 23, no. 10, pp. 787–789, Oct. 2004.

12. M. T. Cushion, M. S. Collins, and M. J. Linke, "Biofilm formation by Pneumocystis spp.," *Eukaryot. Cell*, vol. 8, no. 2, pp. 197–206, Feb. 2009.

13. L. R. Martinez and A. Casadevall, "Cryptococcus neoformans biofilm formation depends on surface support and carbon source and reduces fungal cell susceptibility to heat, cold, and UV light.," *Appl. Environ. Microbiol.*, vol. 73, no. 14, pp. 4592–4601, Jul. 2007.

14. E. Mowat, C. Williams, B. Jones, S. McChlery, and G. Ramage, "The characteristics of Aspergillus fumigatus mycetoma development: is this a biofilm?," *Med. Mycol.*, vol. 47, Suppl 1, no. Supplement I, pp. S120–S126, Jan. 2009.

15. Y. Imamura, J. Chandra, P. K. Mukherjee, A. A. Lattif, L. B. Szczotka-Flynn, E. Pearl- man, J. H. Lass, K. O'Donnell, and M. a Ghannoum, "Fusarium and Candida albicans biofilms on soft contact lenses: model development, influence of lens type, and sus- ceptibility to lens care solutions.," *Antimicrob. Agents Chemother.*, vol. 52, no. 1, pp. 171–182, Jan. 2008.

16. N. S. Pitangui, J. C. O. Sardi, J. F. Silva, T. Benaducci, R. A. Moraes da Silva, G. Rodríguez-Arellanes, M. L. Taylor, M. J. S. Mendes-Giannini, and A. M. Fusco-Almeida, "Adhesion of Histoplasma capsulatumto pneumocytes and biofilm formation on an abiotic surface," *Biofouling*, vol. 28, no. 7, pp. 711–718, 2012.

17. J. D. C. O. Sardi, N. D. S. Pitangui, G. Rodríguez-Arellanes, M. L. Taylor, A. M. Fus- co-Almeida, and M. J. S. Mendes-Giannini, "Highlights in pathogenic fungal bio- films.," *Rev. Iberoam. Micol.*, vol. 31, no. 1, pp. 22–29, 2014.

18. L. E. Davis, G. Cook, and J. W. Costerton, "Biofilm on ventriculoperitoneal shunt tub- ing as a cause of treatment failure in coccidioidal meningitis," *Emerg. Infect. Dis.*, vol. 8, no. 4, pp. 376–379, 2002.

19. R. Singh, M. R. Shivaprakash, and A. Chakrabarti, "Biofilm formation by zygomycetes: quantification, structure and matrix composition," *Microbiology*, vol. 157, no. 9, pp. 2611–2618, 2011.

20. T. P. McCarty and P. G. Pappas, "Invasive Candidiasis," *Infect. Dis. Clin. North Am.*, vol. 00114–2, no. 15, pp. 1445–1456, 2015.

21. C. Seneviratne, L. Jin, and L. Samaranayake, "Biofilm lifestyle of Candida: a mini re- view," *Oral Dis.*, vol. 14, no. 7, pp. 582–590, 2008.

22. L. J. Douglas, "Candida biofilms and their role in infection," *TRENDS Microbiol.*, vol. 11, no. 1, pp. 1–7, 2003.

23. L. a Mermel, M. Allon, E. Bouza, D. E. Craven, P. Flynn, N. P. O'Grady, I. I. Raad, B. J. a Rijnders, R. J. Sherertz, and D. K. Warren, "Clinical practice guidelines for the diagnosis and management of intravascular catheter-related infection: 2009 Update by the Infectious Diseases Society of America.," *Clin. Infect. Dis.*, vol. 49, pp. 1–45, 2009.

24. O. A. Cornely, M. Bassetti, T. Calandra, J. Garbino, B. J. Kullberg, O. Lortholary, W. Meersseman, M. Akova, M. C. Arendrup, S. Arikan-Akdagli, J. Bille, E. Castagnola, M. Cuenca-Estrella, J. P. Donnelly, A. H. Groll, R. Herbrecht, W. W. Hope, H. E. Jen- sen, C. Lass-Flörl, G. Petrikkos, M. D. Richardson, E. Roilides, P. E. Verweij, C. Visco- li, and A. J. Ullmann, "ESCMID guideline for the diagnosis and management

of Candida diseases 2012: Non-neutropenic adult patients," *Clin. Microbiol. Infect.*, vol. 18, no. Suppl.7, pp. 19–37, 2012.

25. M. Nucci, E. Anaissie, R. F. Betts, B. F. Dupont, C. Wu, D. N. Buell, L. Kovanda, and O. Lortholary, "Early removal of central venous catheter in patients with candidemia does not improve outcome: analysis of 842 patients from 2 randomized clinical trials.," *Clin. Infect. Dis.*, vol. 51, no. 3, pp. 295–303, 2010.

26. M. A. Pfaller and D. J. Diekema, "The epidemiology of invasive candidiasis," in *Can- dida and candidiasis.*, 2 ed., R. Calderone, Ed. Washington DC: ASM Press, 2012, pp. 449–480.

27. M. Pfaller and D. Diekema, "Epidemiology of invasive candidiasis: a persistent pub- lic health problem.," *Clin. Microbiol. Rev.*, vol. 20, no. 1, pp. 133–163, Jan. 2007.

28. M. E. Falagas, N. Roussos, and K. Z. Vardakas, "Relative frequency of albicans and the various non-albicans Candida spp among candidemia isolates from inpatients in various parts of the world : a systematic review," *Int. J. Infect. Dis.*, vol. 14, no. 11, pp. e954–e966, 2010.

29. J. D. Sobel, "The emergence of non-albicans Candida species as causes of invasive candidiasis and candidemia.," *Curr. Infect. Dis. Rep.*, vol. 8, no. 6, pp. 427–433, Nov. 2006.

30. M. A. Pfaller, D. R. Andes, D. J. Diekema, D. L. Horn, A. C. Reboli, C. Rotstein, B. Franks, and N. E. Azie, "Epidemiology and Outcomes of Invasive Candidiasis Due to Non-albicans Species of Candida in 2,496 Patients: Data from the Prospective Anti- fungal Therapy (PATH) Registry 2004–2008.," *PLoS One*, vol. 9, no. 7, p. e101510, Jan. 2014.

31. A. A. Cleveland, M. M. Farley, L. H. Harrison, B. Stein, R. Hollick, S. R. Lockhart, S. S. Magill, G. Derado, B. J. Park, and T. M. Chiller, "Changes in incidence and antifun- gal drug resistance in candidemia: results from population-based laboratory surveil- lance in Atlanta and Baltimore, 2008–2011.," *Clin. Infect. Dis.*, vol. 55, no. 10, pp. 1352– 1361, Nov. 2012.

32. M. A. Pfaller, S. A. Messer, L. N. Woosley, R. N. Jones, and M. Castanheira, "Echinocandin and triazole antifungal susceptibility profiles for clinical opportunistic yeast and mold isolates collected from 2010 to 2011: application of new CLSI clinical break- points and epidemiological cutoff values for characterization of geographic," *J. Clin. Microbiol.*, vol. 51, no. 8, pp. 2571–2581, Aug. 2013.

33. F. Chapeland-Leclerc, C. Hennequin, N. Papon, T. Noël, A. Girard, G. Socié, P. Ribaud, and C. Lacroix, "Acquisition of flucytosine, azole, and caspofungin resistance in Candida glabrata bloodstream isolates serially obtained from a hematopoietic stem cell transplant recipient," *Antimicrob. Agents Chemother.*, vol. 54, no. 3, pp. 1360–1362, 2010.

34. G. Ramage, B. L. Wickes, and J. L. Lopez-Ribot, "A seed and feed model for the for- mation of Candida albicans biofilms under flow conditions using an improved modi- fied Robbins device," *Rev. Iberoam. Micol.*, vol. 25, pp. 37–40, 2008.

35. S. Silva, M. Henriques, A. Martins, R. Oliveira, D. Williams, and J. Azeredo, "Biofilms of non-Candida albicans Candida species: quantification, structure and matrix composition," *Med. Mycol.*, vol. 47, no. 7, pp. 681–689, 2009.

36. D. R. Andes, J. Nett, P. Oschel, R. Albrecht, K. Marchillo, and A. Pitula, "Development and characterization of an in vivo central venous catheter Candida albicans bio- film model," *Infect. Immun.*, vol. 72, no. 10, pp. 6023–6031, 2004.

37. J. E. Nett, K. Marchillo, C. A. Spiegel, and D. R. Andes, "Development and validation of an in vivo Candida albicans biofilm denture model," *Infect. Immun.*, vol. 78, no. 9, pp. 3650–3659, 2010.

38. M. Ricicova, S. Kucharikova, H. Tournu, J. Hendrix, H. Bujdakova, J. Van Eldere, K. Lagrou, and P. Van Dijck, "Candida albicans biofilm formation in a new in vivo rat model," *Microbiology*, vol. 156, no. 3, pp. 909–919, 2010.

39. J. Crump and P. Collignon, "Intravascular catheter-associated infections.," *Eur. J. Clin. Microbiol. Infect. Dis.*, vol. 19, no. 1, pp. 1–8, Jan. 2000.

40. H. Wisplinghoff, T. Bischoff, S. M. Tallent, H. Seifert, R. P. Wenzel, and M. B. Edmond, "Nosocomial bloodstream infections in US hospitals: analysis of 24,179 cases from a prospective nationwide surveillance study.," *Clin. Infect. Dis.*, vol. 39, no. 3, pp. 309–17, Aug. 2004.

41. M. H. Nguyen, J. E. J. Peacock, D. C. Tanner, A. J. Morris, M. L. Nguyen, D. R. Snyd- man, M. M. Wagener, and V. L. Yu, "Therapeutic approaches in patients with candi- demia: Evaluation in a multicenter, prospective, observational study," *Arch. Intern. Med.*, vol. 155, no. 22, pp. 2429–2435, 1995.

42. R. Rajendran, L. Sherry, C. J. Nile, A. Sherriff, E. M. Johnson, M. F. Hanson, C. Williams, C. A. Munro, B. J. Jones, and G. Ramage, "Biofilm formation is a risk factor for mortality in patients with Candida albicans bloodstream infection—Scotland, 2012– 2013," *Clin. Microbiol. Infect.*, pp. 1–7, 2015.

43. J. Chandra, D. M. Kuhn, P. K. Mukherjee, L. L. Hoyer, T. McCormick, and M. A. Ghannoum, "Biofilm formation by the fungal pathogen Candida albicans: development, architecture, and drug resistance," *J. Bacteriol.*, vol. 183, no. 18, pp. 5385–5394, 2001.

44. G. Ramage, K. VandeWalle, B. L. Wickes, and J. L. Lopez-Ribot, "Characteristics of biofilm formation by Candida albicans," *Rev. Iberoam. Micol.*, vol. 18, pp. 163–170, 2001.

45. G. Ramage, S. P. Saville, D. P. Thomas, and J. L. J. L. Lopez-Ribot, "Candida biofilms: an update," *Eukaryot. Cell*, vol. 4, no. 4, pp. 633–638, 2005.

46. C. Loussert, C. Schmitt, M.-C. Prévost, V. Balloy, E. Fadel, B. Philippe, C. Kauffmann-Lacroix, J. P. Latgé, and A. Beauvais, "In vivo biofilm composition of Aspergillus fu- migatus," *Cell. Microbiol.*, vol. 12, no. 3, pp. 405–410, 2010.

47. G. Ramage, R. Rajendran, M. Gutierrez-Correa, B. Jones, and C. Williams, "Aspergillus biofilms: clinical and industrial significance," *FEMS Microbiol. Lett.*, vol. 324, pp. 89–97, 2011.

48. S. G. Filler and D. C. Sheppard, "Fungal invasion of normally non-phagocytic host cells," *PLoS Pathog.*, vol. 2, no. 12, p. e129, 2006.

49. A. Beauvais, C. Schmidt, S. Guadagnini, P. Roux, E. Perret, C. Henry, S. Paris, A. Mallet, M.-C. Prévost, and J. P. Latgé, "An extracellular matrix glues together the aer- ial-grown hyphae of Aspergillus fumigatus," *Cell. Microbiol.*, vol. 9, no. 6, pp. 1588– 1600, 2007.

50. E. Mowat, J. Butcher, S. Lang, C. Williams, and G. Ramage, "Development of a sim- ple model for studying the effects of antifungal agents on multicellular communities of Aspergillus fumigatus.," *J. Med. Microbiol.*, vol. 56, no. Pt 9, pp. 1205–1212, Sep. 2007.

51. M. J. Seidler, S. Salvenmoser, and F.-M. C. Müller, "Aspergillus fumigatus forms bio- films with reduced antifungal drug susceptibility on bronchial epithelial cells.," *Anti- microb. Agents Chemother.*, vol. 52, no. 11, pp. 4130–4136, Nov. 2008.

52. D. Singhal, L. Baker, P. Wormald, and Tan L, "Aspergillus fumigatus biofilm on pri- mary human sinonasal epithelial culture.," *Am. J. Rhinol. Allergy*, vol. 25, no. 4, pp. 219–225, 2011.

53. W. Escande, G. Fayad, T. Modine, E. Verbrugge, M. Koussa, E. Senneville, and O. Le- roy, "Culture of a prosthetic valve excised for streptococcal endocarditis positive for Aspergillus fumigatus 20 years after previous A fumigatus endocarditis.," *Ann. Thor- ac. Surg.*, vol. 91, no. 6, pp. e92–e93, Jun. 2011.

54. T. K. Jeloka, S. Shrividya, and G. Wagholikar, "Catheter outflow obstruction due to an aspergilloma.," *Perit. Dial. Int.*, vol. 31, no. 2, pp. 211–212, 2011.

55. G. Ramage, R. Rajendran, L. Sherry, and C. Williams, "Fungal biofilm resistance," *Int. J. Microbiol.*, vol. 2012, pp. 1–14, 2012.

56. F.-M. C. Müller, M. Seidler, and A. Beauvais, "Aspergillus fumigatus biofilms in the clinical setting.," *Med. Mycol.*, vol. 49 Suppl 1, no. April, pp. S96–S100, Apr. 2011.

57. B. Fiori, B. Posteraro, R. Torelli, M. Tumbarello, D. S. Perlin, G. Fadda, and M. San- guinetti, "In vitro activities of anidulafungin and other antifungal agents against bio- films formed by clinical isolates of different Candida and Aspergillus species.," *Antimicrob. Agents Chemother.*, vol. 55, no. 6, pp. 3031–3035, Jun. 2011.

58. F. P. Gullo, S. A. Rossi, J. D. C. O. Sardi, V. L. I. Teodoro, M. J. S. Mendes-Giannini, and A. M. Fusco-Almeida, "Cryptococcosis: epidemiology, fungal resistance, and new alternatives for treatment.," *Eur. J. Clin. Microbiol. Infect. Dis.*, vol. 32, no. 11, pp. 1377–1391, Nov. 2013.

59. B. J. Park, K. a Wannemuehler, B. J. Marston, N. Govender, P. G. Pappas, and T. M. Chiller, "Estimation of the current global burden of cryptococcal meningitis among persons living with HIV/AIDS.," *AIDS*, vol. 23, no. 4, pp. 525–530, Feb. 2009.

60. L. R. Martinez and A. Casadevall, "Specific antibody can prevent fungal biofilm for- mation and this effect correlates with protective efficacy," vol. 73, no. 10, pp. 6350– 6362, 2005.

61. M. Nucci and E. Anaissie, "Fusarium infections in immunocompromised patients.," *Clin. Microbiol. Rev.*, vol. 20, no. 4, pp. 695–704, Oct. 2007.

62. P. K. Mukherjee, J. Chandra, C. Yu, Y. Sun, E. Pearlman, and M. A. Ghannoum, "Characterization of Fusarium keratitis outbreak isolates: contribution of biofilms to antimicrobial resistance and pathogenesis," *Investig. Ophthalmol. Vis. Sci.*, vol. 53, no. 8, pp. 4450–4457, 2012.

63. C. N. Burkhart, C. G. Burkhart, and A. K. Gupta, "Dermatophytoma: recalcitrance to treatment because of existence of fungal biofilm," *J. Am. Acad. Dermatol.*, vol. 47, no. 4, pp. 629–631, Oct. 2002.

64. A. G. Nusbaum, R. S. Kirsner, and C. A. Charles, "Biofilms in dermatology," *Skin Therapy Lett.*, vol. 17, no. 7, pp. 1–5, 2012.

65. Y. Sun, J. Chandra, P. Mukherjee, L. Szczotka-Flynn, M. A. Ghannoum, and E. Pearl- man, "A murine model of contact lens-associated Fusarium keratitis.," *Investig. Oph- thalmol. Vis. Sci.*, vol. 51, no. 3, pp. 1511–1516, 2010.

66. T. V. M. Vila, N. S. Quintanilha, and S. Rozental, "Miltefosine is effective against Candida albicans and Fusarium oxysporum nail biofilms in vitro," *J. Med. Microbiol.*, pp. 1–14, 2015.

67. T. V. M. Vila, S. Rozental, and C. M. D. de S. Guimarães, "A new model of in vitro fungal biofilms formed on human nail fragments allows reliable testing of laser and light therapies against onychomycosis," *Lasers Med. Sci.*, vol. 30, no. 3, pp. 1031–1039, 2015.

68. F. C. Odds, A. J. P. Brown, and N. A. R. Gow, "Antifungal agents: mechanisms of ac- tion," *Trends Microbiol.*, vol. 11, no. 6, pp. 272–279, Jun. 2003.

69. R. S. Shapiro, N. Robbins, and L. E. Cowen, "Regulatory circuitry governing fungal development, drug resistance, and disease.," *Microbiol. Mol. Biol. Rev.*, vol. 75, no. 2, pp. 213–267, Jun. 2011.

70. J. Chandra, P. K. Mukherjee, S. D. Leidich, F. F. Faddoul, L. L. Hoyer, L. J. Douglas, and M. A. Ghannoum, "Antifungal resistance of candidal biofilms formed on den- ture acrylic in vitro," *J. Dent. Res.*, vol. 80, no. 3, pp. 903–908, 2001.

71. D. M. Kuhn, T. George, J. Chandra, P. K. Mukherjee, and M. A. Ghannoum, "Anti- fungal susceptibility of candida biofilms: unique efficacy of amphotericin B lipid for- mulations and echinocandins," *Antimicrob. Agents Chemother.*, vol. 46, pp. 1773–1780, 2002.

72. A. M. Tortorano, "The European Confederation of Medical Mycology (ECMM) sur- vey of candidaemia in Italy: in vitro susceptibility of 375 Candida albicans isolates and biofilm production," *J. Antimicrob. Chemother.*, vol. 56, no. 4, pp. 777–779, 2005.

73. K. Ajesh and K. Sreejith, "Cryptococcus laurentii biofilms: structure, development and antifungal drug resistance.," *Mycopathologia*, vol. 174, no. 5–6, pp. 409–419, Dec. 2012.

74. L. R. Martinez and A. Casadevall, "Susceptibility of Cryptococcus neoformans bio- films to antifungal agents in vitro," *Antimicrob. Agents Chemother.*, vol. 50, no. 3, pp. 1021–1033, 2006.

75. E. Mowat, S. Lang, C. Williams, E. McCulloch, B. Jones, and G. Ramage, "Phase-de- pendent antifungal activity against Aspergillus fumigatus developing multicellular

filamentous biofilms.," *J. Antimicrob. Chemother.*, vol. 62, no. 6, pp. 1281–1284, Dec. 2008.

76. X. Zhang, X. Sun, Z. Wang, Y. Zhang, and W. Hou, "Keratitis-associated fungi form biofilms with reduced antifungal drug susceptibility," *Investig. Ophthalmol. & Vis. Sci.*, vol. 53, no. 12, pp. 7774–7778, 2012.

77. G. Ramage, S. Bachmann, T. F. Patterson, B. L. Wickes, and J. L. López-Ribot, "Investigation of multidrug efflux pumps in relation to fluconazole resistance in Candida albicans biofilms," *J. Antimicrob. Chemother.*, vol. 49, no. 6, pp. 973–980, 2002.

78. P. Perumal, S. Mekala, and W. L. Chaffin, "Role for cell density in antifungal drug resistance in Candida albicans biofilms," *Antimicrob. Agents Chemother.*, vol. 51, no. 7, pp. 2454–2463, 2007.

79. D. M. Kuhn, J. Chandra, P. K. Mukherjee, and M. A. Ghannoum, "Mechanism of flu- conazole resistance in Candida albicans biofilms: phase-specific role of efflux pumps and membrane sterols," *Infect. Immun.*, vol. 71, no. 8, pp. 878–888, 2002.

80. A. A. Lattif, P. K. Mukherjee, J. Chandra, M. R. Roth, R. Welti, M. Rouabhia, and M. A. Ghannoum, "Lipidomics of Candida albicans biofilms reveals phase-dependent production of phospholipid molecular classes and role for lipid rafts in biofilm formation," *Microbiology*, vol. 157, no. 11, pp. 3232–3242, 2011.

81. P. K. Mukherjee and J. Chandra, "Candida biofilm resistance," *Drug Resist. Updat.*, vol. 7, no. 4–5, pp. 301–309, 2004.

82. H. Nailis, S. Kucharíková, M. Řičicová, P. Van Dijck, D. Deforce, H. Nelis, and T. Coenye, "Real-time PCR expression profiling of genes encoding potential virulence factors in Candida albicans biofilms: identification of model-dependent and -independent gene expression," *BMC Microbiol.*, vol. 10, no. 1, p. 114, 2010.

83. S. Borecká-Melkusová, G. P. Moran, D. J. Sullivan, S. Kucharíková, D. Chorvát Jr, and H. Bujdáková, "The expression of genes involved in the ergosterol biosynthesis pathway in Candida albicansand Candida dubliniensis biofilms exposed to fluconazole," *Mycoses*, vol. 52, no. 2, pp. 118–128, 2009.

84. T. Rossignol, C. Ding, A. Guida, C. D'Enfert, D. G. Higgins, and G. Butler, "Correlation between biofilm formation and the hypoxic response in Candida parapsilosis," *Eukaryot. Cell*, vol. 8, no. 4, pp. 550–559, 2009.

85. J. E. Nett, A. J. Lepak, K. Marchillo, and D. R. Andes, "Time course global gene expression analysis of an in vivo Candida biofilm," *J. Infect. Dis.*, vol. 20023, no. 2, pp. 307–313, 2009.

86. J. W. Song, J. H. Shin, S. J. Kee, S. H. Kim, M. G. Shin, S. P. Suh, and D. W. Ryang, "Expression of CgCDR1, CgCDR2, and CgERG11 in Candida glabrata biofilms formed by bloodstream isolates.," *Med. Mycol.*, vol. 47, no. 5, pp. 545–548, 2009.

87. C. Mateus, S. a. Crow, and D. G. Ahearn, "Adherence of Candida albicans to silicone induces immediate enhanced tolerance to fluconazole," *Antimicrob. Agents Chemother.*, vol. 48, no. 9, pp. 3358–3366, 2004.

88. A. Lepak, J. Nett, L. Lincoln, K. Marchillo, and D. Andes, "Time course of microbiologic outcome and gene expression in Candida albicans during and following in vitro and in vivo exposure to fluconazole time course of microbiologic outcome and gene expression in Candida albicans during and following in vitro and in vivo," *Antimicrob. Agents Chemother.*, vol. 50, no. 4, pp. 1311–1319, 2006.

89. M. A. Al-Fattani, "Biofilm matrix of Candida albicans and Candida tropicalis: chemi- cal composition and role in drug resistance," *J. Med. Microbiol.*, vol. 55, no. 8, pp. 999– 1008, 2006.

90. G. S. Baillie and L. J. Douglas, "Matrix Polymers of Candida biofilms and their possible role in biofilm resistance to antifungal agents," *J. Antimicrob. Chemother.*, vol. 46, pp. 397–403, 2000.

91. R. Zarnowski, W. M. Westler, G. A. Lacmbouh, J. M. Marita, J. R. Bothe, J. Bernhardt, A. S. Lounes-Hadj, J. Fontaine, H. Sanchez, R. D. Hatfield, J. M. Ntambi, J. E. Nett, A. P. Mitchell, and D. R. Andes, "Novel entries in a fungal biofilm matrix encyclopedia," *MBio*, vol. 5, no. 4, 2014.

92. J. Nett, L. Lincoln, K. Marchillo, R. Massey, K. Holoyda, B. Hoff, M. VanHandel, and D. Andes, "Putative role of -1,3 glucans in Candida albicans biofilm resistance," *Anti- microb. Agents Chemother.*, vol. 51, no. 2, pp. 510–520, 2007.

93. J. E. Nett, K. Crawford, K. Marchillo, and D. R. Andes, "Role of Fks1p and matrix glucan in Candida albicans biofilm resistance to an echinocandin, pyrimidine, and polyene," *Antimicrob. Agents Chemother.*, vol. 54, no. 8, pp. 3505–3508, 2010.

94. J. E. Nett, H. Sanchez, M. T. Cain, and D. R. Andes, "Genetic basis of Candida biofilm resistance due to drug-sequestering matrix glucan," *J. Infect. Dis.*, vol. 202, no. 1, pp. 171–175, 2010.

95. G. Vediyappan, T. Rossignol, and C. D. Enfert, "Interaction of Candida albicans bio- films with antifungals: transcriptional response and binding of antifungals to beta- glucans," *Antimicrob. Agents Chemother.*, vol. 54, no. 5, pp. 2096–2111, 2010.

96. M. Martins, P. Uppuluri, D. Thomas, I. Cleary, M. Henriques, J. Lopez-Ribot, and R. Oliveira, "Presence of extracellular DNA in the Candida albicans biofilm matrix and its contribution to biofilms.," *Mycopathologia*, vol. 169, no. 5, pp. 323–331, 2010.

97. M. Martins, M. Henriques, J. L. Lopez-Ribot, and R. Oliveira, "Addition of DNase improves the in vitro activity of antifungal drugs against Candida albicans biofilms," *Mycoses*, vol. 55, no. 1, pp. 80–85, 2012.

98. R. Rajendran, C. Williams, D. F. Lappin, O. Millington, M. Martins, and G. Ramage, "Extracellular DNA release acts as an antifungal resistance mechanism in mature As- pergillus fumigatus biofilms," *Eukaryot. Cell*, vol. 12, no. May, pp. 420–429, 2013.

99. C. J. Nobile, J. E. Nett, A. D. Hernday, O. R. Homann, J.-S. Deneault, A. Nantel, D. R. Andes, A. D. Johnson, and A. P. Mitchell, "Biofilm matrix regulation by Candida al- bicans Zap1," *PLoS Biol.*, vol. 7, no. 6, p. e1000133, 2009.

100. K. Lewis, "Persister cells.," *Annu. Rev. Microbiol.*, vol. 64, pp. 357–372, 2010.

101. R. S. Al-Dhaheri and L. J. Douglas, "Absence of amphotericin B-tolerant persister cells in biofilms of some Candida species," *Antimicrob. Agents Chemother.*, vol. 52, no. 5, pp. 1884–1887, 2008.

102. C. G. Pierce, A. K. Chaturvedi, A. L. Lazzell, A. T. Powell, S. P. Saville, S. F. McHar- dy, and J. L. Lopez-Ribot, "A novel small molecule inhibitor of Candida albicans bio- film formation, filamentation and virulence with low potential for the development of resistance," *npj Biofilms Microbiomes*, vol. 1, no. April, p. 15012, 2015.

103. P. Uppuluri, A. K. Chaturvedi, A. Srinivasan, M. Banerjee, A. K. Ramasubramaniam, J. R. Köhler, D. Kadosh, and J. L. Lopez-Ribot, "Dispersion as an important step in the Candida albicans biofilm developmental cycle," *PLoS Pathog.*, vol. 6, no. 3, p. e1000828, 2010.

104. J. H. Rex, J. E. Bennett, A. M. Sugar, P. G. Pappas, J. Serody, J. E. Edwards, R. G. Washburn, J. H. Rex, J. E. Bennett, A. M. Sugar, P. G. Pappas, J. Serody, J. E. Ed- wards, and R. G. Washburn, "Intravascular catheter exchange and duration of candi- demia," *Clin. Infect. Dis.*, vol. 21, pp. 994–996, 1995.

105. D. R. Andes, N. Safdar, J. W. Baddley, G. Playford, A. C. Reboli, J. H. Rex, J. D. Sobel, P. G. Pappas, and B. J. Kullberg, "Impact of treatment strategy on outcomes in pa- tients with candidemia and other forms of invasive candidiasis: A patient-level quan- titative review of randomized trials," *Clin. Infect. Dis.*, vol. 54, no. 8, pp. 1110–1122, 2012.

106. L. A. Mermel, B. M. Farr, R. J. Sherertz, I. I. Raad, N. O. Grady, J. S. Harris, and D. E. Craven, "Guidelines for the manegment of intravascular catheter-related infections," *Clin. Infect. Dis.*, vol. 32, pp. 1249–1272, 2001.

107. D. Tacke, P. Koehler, B. Markiefka, and O. A. Cornely, "Our 2014 approach to mucor- mycosis," *Mycoses*, vol. 57, pp. 519–524, 2014.

Chapter-2

Infection Mechanisms and Colonization Patterns of Fungi Associated with Soybean

1. Introduction

There is evidence based on molecular predictions to suggest that fungi evolved over one billion years ago [1]. Their associations with plants most likely began over 425 million years

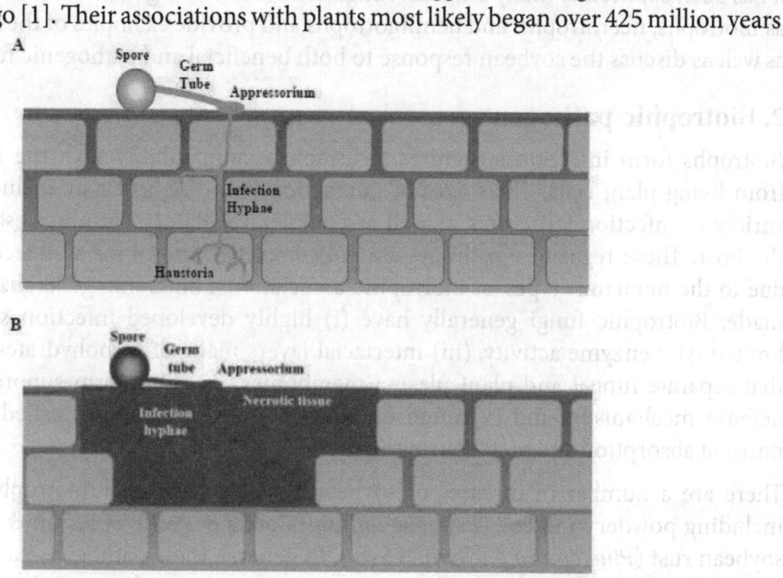

Figure 1. Generalized diagram of infection and colonization patterns of a) biotrophic pathogens; b) necrotrophic patho- gens; and c) biotrophic mutualists.

ago when plants began to colonize land [2]. There are well over 10,000 fungal species known to be associated with living plants and they range from beneficial to pathogenic. Fungal associations with plants can be characterized based on the different ways fungi infect and colonize host plants as well as how fungi use plants as a food source. Some fungal species cannot live without a host plant and are referred to as obligate biotrophs. Obligate biotrophs do not have a saprophytic independent life stage , normally only grow on living plants, and cannot be cultured. Fungi that derive their nutrition from dead tissue are referred to as necrotrophs. These have mechanisms to kill the host tissue prior to infection and colonization and tend to be easily cultured. Other fungal species that may have both biotrophic and necrotrophic phases in their life cycle are referred to as hemibiotrophs. We have chosen soybean [*Glycine max* (L.) Merr.] as the host to demonstrate these different associations that fungi have with plants.

Worldwide, soybean is the most important legume crop and the fourth leading crop in area of production [3]. There are many fungi associated with soybean plants and there are more than 40 named diseases [3]. Since soybean is planted in most cropping areas of the world, it has been exposed to many different fungi. We describe fungal interactions with soybean as biotrophs, necrotrophs, and hemibiotrophs and provide examples of these interactions as well as discuss the soybean response to both beneficial and pathogenic fungi.

2. Biotrophic pathogens

Biotrophs form infection structures to establish compatibility with the host and feed from living plant cells. The stages of interaction with the host may include a number/ variety of infection structures as well as complex regulatory pathways established with the host. These regulatory pathways are not completely known for all infection structures due to the numerous types of biotrophic associations, but some generalizations can be made. Biotrophic fungi generally have (i) highly developed infection structures, (ii) limited lytic enzyme activity, (iii) interfacial layers made of carbohydrates and proteins that separate fungal and plant plasma membranes, (iv) long-term suppression of host defense mechanisms and (v) often develop specialized structures called haustoria for nutrient absorption and metabolism (Figure 1A) [4].

There are a number of diseases of soybean that are caused by biotrophic pathogens, including powdery mildew (*Erysiphe diffusa* (Cooke & Peck) U. Braun & S. Takam and soybean rust (*Phakopsora pachyrhizi* Syd.). In general, these pathogens have greater host specificity than pathogens classified as necrotrophs and often have developed more strain specificity with specific soybean genotypes, leading to gene-for-gene interactions and race development in the pathogen [3]. The example we are using as an obligate biotroph is *P. pachyrhizi*.

Like most rust fungi, *P. pachyrhizi* produces copious amounts of wind-blown spores called urediniospores. When these urediniospores land on a soybean leaf, they germinate and each forms a single germ tube of about 3 μm in width and of varying length from 100 μm or more before terminating into an appressorium that is approximately the same size as a uredinio- spore. The appressorium serves as an attachment structure at the site of penetration (Figure 2) [5, 6]. The appressorium develops an appressorial cone that initiates the penetration into the epidermal cell by turgor pressure independent of melanin accumulation [7]. The penetration hyphae grow through the epidermal cell and intercellular space, first forming primary invading hyphae, and then secondary hyphae populating intercellular spaces. In compatible interactions, primary haustoria form in the mesophyll cells, and within 12 days the formation of secondary haustoria occurs. A domed-shaped eruption occurs in the epidermis sometimes as early as 8 days after infection to form uredinia that produce urediniospores, causing the leaf to have a rusty-tan appearance. These urediniospores are wind-blown and serve as the source of inoculum for new infections.

When the soybean defense system is activated, an incompatible interaction occurs. In this case, fungal hyphae in the mesophyll cells become less common as mesophyll cells become necrotic (Figure 3). In addition, organized haustoria and uredinia do not develop, resulting in no sporulation [5, 8]. The genetics of the incompatible interaction have been studied and six loci (*Rpp1* to *Rpp6*) are known to have dominant genes that segregate independently [9]. Some of these genes provide a strong or absolute resistance to certain isolates of the fungus while other genes show an incomplete resistance to certain isolates, which allows for sporulation although usually reduced compared to a compatible interaction [10].

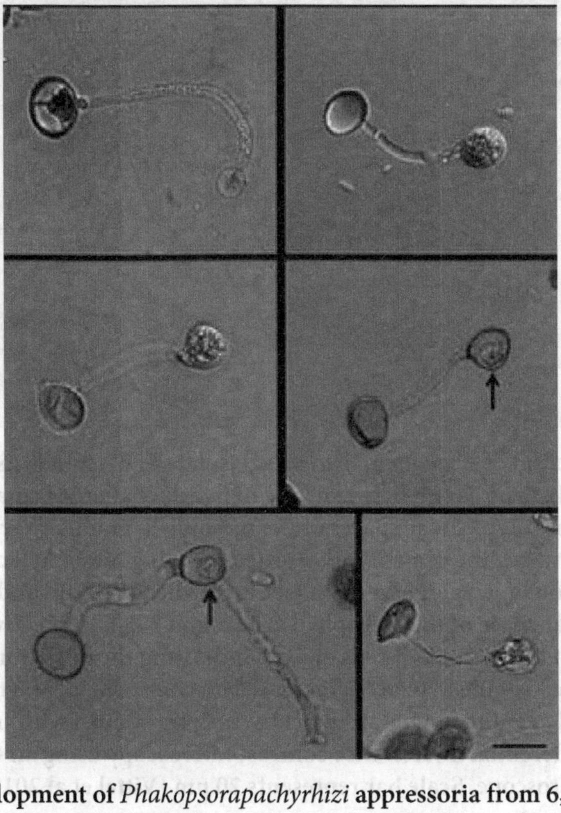

Figure 2. Development of *Phakopsorapachyrhizi* appressoria from 6, 8, 12, 24 to 48, respectively (top left to bottom left), and collapse of the spore and appressorium due to plasmolysis appressorium (bottom right). Arrows point to appres- soria. Bar on bottom right represents 25 μm (Chang et al. 2014).

A study that combined both histological development and fungal DNA quantification during infection and colonization [5] showed that early events of spore germination, appressorium formation, and fungal penetration of the epidermis occurred within 24 hours post-inoculation among the all tested soybean genotypes; differences in infection among genotypes were evident once the hyphae penetrated into the intercellular spaces between the mesophyll cells. For example, at 2 days after inoculation, the compatible soybean genotype Williams 82 had a higher percentage of hyphae in the mesophyll tissue than the incompatible soybean genotype, and the percentage of interaction sites with mesophyll cell death was higher in the most incompatible genotype starting at 3 days after inoculation. This study also reported a positive correlation between quantities of hyphae in the mesophyll cells and fungal DNA and demon- strates that an incompatible soybean-*P. pachyrhizi* interaction results in restricted hyphal development in mesophyll cell tissue, likely due to hypersensitive apoptosis [5].

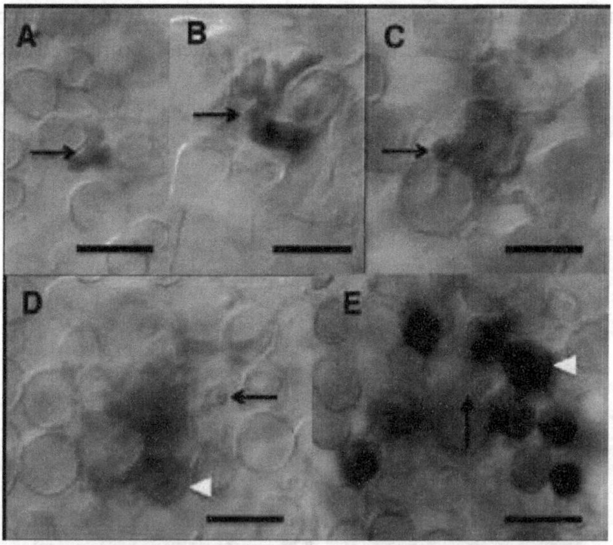

Figure 3. Infection of *Phakospora pachyrhizi* isolate FL07-1 on resistant soybean genotype PI 224268 observed between one to five days after inoculation (dai). The fungus entered the intercellular spaces of the mesophyll, gradually grew (see arrows) until 3 dai after which the growth was arrested by dying mesophyll cells, visualized by their retention of trypan blue stain (see arrow-heads). A, B, and C, Hyphae in the intercellular space of the mesophyll at 1, 2, and 3 dai, respectively. D, Slight discoloration of the mesophyll cells at 4 dai indicating the initiation of defense to arrest the spread of the hyphae. E, Increased discoloration of several mesophyll cells around and near the fungus at 5 dai. Leaves were stained with trypan blue and observed with differential interference contrast microscopy using an Olympus BX 51 microscope. Scale bar represents 20 μm (Vittal et al 2014).

In another study, soybean transcriptome changes during soybean rust infection showed an early burst of gene expression at 12 h in both compatible and incompatible interactions [11]. This corresponds to the progression of urediniospore germination and appressorial formation, showing that these infection events induce dramatic changes in plant gene expression. Gene expression was much less from 1 to 2 days after inoculation, the period when hyphae entered into the mesophyll cells, with major differences in gene expression between compatible and incompatible genotypes noted at 3 days after inoculation, corresponding to haustoria forma- tion and suggesting that genes expressed at 3 days after inoculation were involved in resistance gene-mediated defense responses [11]. In another study, the inoculation with a *P. pachyrhizi* isolate and an incompatible soybean genotype showed an up-regulation of peroxidases with an oxidative burst in infected cells that triggered programmed cell death or apoptosis [12].

The soybean rust pathogen has a wide host range of legumes specifically those in the subfamily Papilionoideae [13]. Although it is not known if these hosts have similar infection and colonization processes to soybean, some studies have focused on the interaction of this fungus with non-hosts. For example, it was shown that plants of *Arabidopsis thaliana* (L.) Heynh. inoculated with *P. pachyrhizi* urediniospores germinated and penetrated the epidermal cells, but did not grow either inter- or intracellularly into the mesophyll tissue layer [14]. In *Hordeum vulgare* L., it was shown that fungal urediniospores germinated, but mostly failed to penetrate through the cuticle and the epidermal cells; however, when the fungus successfully penetrated, the epidermal cells died and colonization was arrested by the hypersensitive collapse of the mesophyll cells [15]. In a study using *Medicago truncatula* Gärtner as a non-host, urediniospores germinated to form appressoria and the fungus penetrated into the epidermal cells, but the fungus did not sporulate [16]. Furthermore, this study showed that genes that produced mecicarpan and chlorophyll catabolism were induced by *M. truncatula* soon after infection.

3. Necrotrophs

Necrotrophic pathogens survive by killing plant tissue to obtain nutrients and advance through colonization of dead tissue. Historically, necrotrophic infection was not considered a specific host-pathogen interaction because these fungi often excrete phytotoxins and cell wall degrading enzymes to kill host tissue with no reciprocating response from the host. Some have characteristics of both biotrophs and necrotrophs and are classified as hemibiotrophs, such as in the genus *Colletotrichum*. Hemibiotrophs will be covered in the next section.

Initial stages of infection by necrotrophs often start when conidia germinate and form infection hyphae that directly penetrate or the fungus develops appressoria that form penetration pegs to penetrate the epidermis (Figure 1B) [17]. Appressoria excrete a number of lytic enzymes including oxidases, cutinases, and lipases to degrade the plant cuticle and wax layers. The penetration pegs breach through the compromised epidermal cells and secrete various cell wall degrading enzymes (CWDEs) to kill cells and progress through the plant causing necrotic lesions and, in some cases, plant death. These enzymes are used by necrotrophs to break down cells and gain access to host nutrients and carbon. *Macrophomina phaseolina* (Tassi) Goid. isolates are known to produce an abundance of CWDEs including hydrolytic and lignin degradation enzymes [18]. Necrotrophs will also commonly secrete phytotoxins to facilitate colonization. These can be broad spectrum, such as (-)-botryodiplodin and phaseolinone produced by *M. phaseolina* (Radwan et al. 2014), or host-specific, such as tomato-specific AAL-toxins produced by *Alternaria alternate* f. sp. *lycopersici* Keissl. [19]. Botrydial, a toxin produced by *Botrytis cinerea* Pers., will cause chlorosis, cell collapse, and aids in fungal penetration of tissues [20]. Botrydial, similar to other necrotrophic phytotoxins, may contribute to the host unspecificity of the fungus [20].

There are several necrotrophs that infect and colonize soybean. Root pathogens *Cadophora gregata* (Allington & D.W. Chamb.) T.C. Harr. & McNew (brown stem rot), *Fusarium virguli- forme* O'Donnell & T. Aoki (sudden death syndrome), *M. phaseolina*, and *Rhizoctonia solani* J.G. Kühn (Rhizoctonia damping off and root rot) among others are common to soybean are ubiquitous and have a broad host range [3]. Necrotrophs can also be above ground pathogens like *Cercospora kikuchii* T. Matsumoto & Tomoy. (purple seed stain), *Cercospora sojina* K. Hara (frogeye leaf spot), and *Septoria glycines* Hemmi (brown spot) that are also ubiquitous [3].

In some necrotrophs, there is a sophisticated microbe-host interaction where the pathogen takes over host defenses. *Sclerotinia sclerotiorum* (Lib.) de Bary has previously been considered a primitive fungal pathogen, but actually has a sophisticated interaction with the host plant. The pathogen survives as sclerotia in the soil that germinate to produce mycelia or produce apothecia that contain ascospores. Apothecia are sexual fruiting bodies produced on a stipe that emerge from sclerotia [21]. When there is a relative change in humidity or a physical disturbance, mature apothecia forcibly eject ascospores in a puff. The majority of ascospores will remain in the field where they were produced; however, they can also be carried several kilometers in air currents [21].

The infection process begins when ascospores germinate on host tissue. To germinate, ascospores need free water and an exogenous nutrient source to grow like that of senescing flower petals. As the fungus colonizes these petals, it excretes oxalic acid and kills cells for further growth on the peduncle or onto a petiole or a leaf [21]. A study compared the infection patterns of *S. sclerotiorum* on cotyledons of a susceptible and a resistance cultivar of *Brassica napus* L. and showed that fungal hyphae grew along the surface of the plant and directly penetrated the cuticle by forming an appressorium or indirectly through stomata [22]. This process differed between susceptible and resistant genotypes after 24 hours post-inoculation when it was observed that there were longer hyphal strands on the susceptible genotype followed by dichotomous branching of terminal hyphae and appressorium development. In the resistant genotype, the fungus produced swollen hyphal apices and hyphal progression was restricted to the epidermal layer while the fungus continued to grow in the susceptible genotype and was able to colonize the mesophyll cells causing cell death. Hyphal cell collapse was also observed only on the resistant genotype. These contrasting interactions during infection and colonization between the susceptible and resistant genotypes indicates signaling between fungal growth and the host occurred at first contact or soon after and caused the fungus to be suppressed in the resistant genotype and not suppressed in the susceptible genotype as appressoria formed and the plant became colonized.

Gene expression profiling was used to characterize gene expression differences of a highly susceptible soybean cultivar and a partially resistant cultivar during the early stages of *S. sclerotiorum* infection [23]. Over 100 genes of interest including genes regulating cell wall composition, signaling pathways, and anthocyanin and anthocyanidin synthesis were reported. The fungus also has over 100 genes that encode for CWDEs and genes for the production of the phytotoxin sclerin [17]. In addition, the fungus produces and secretes oxalic acid [24, 25], which in itself is toxic to the plant, but also modulates a number of plant phys- iological processes that allow the fungus to invade unrecognized by the host. This phenom- enon is similar to biotrophic pathogens, and newer information has stated that *S. sclerotiorum* is a hemibiotroph [26]. The main role of oxalic acid, or oxalate, in pathogenicity is the manipulation of host defense mechanisms. Initially, oxalate renders the reactive oxygen species (ROS) response ineffective. The oxidative burst and callose deposition is an initial defense mechanism to inhibit further colonization. In later stages, the fungus triggers the ROS mechanisms to initiate programmed cell death [25]. The fungal hyphae are then able to colonize the dead tissue and continue this process down the petiole and through the stem of the soybean eventually killing the entire plant (Figure 4). Oxalic acid is also involved in pH reduction, guard cell regulation, and chelation of calcium [25].

Figure 4. Necrotic tissue and water-soaking on a soybean leaf infected with Sclerotinia sclerotiorum.

4. Hemibiotrophs

There are some pathogens that deploy both biotrophic and necrotrophic mechanisms for infection and colonization and are considered by some to be the best-armed phytopathogens [27]. Commonly, a hemibiotrophic fungus has a biotrophic phase where it secures a relation- ship with the host plant before switching to a necrotrophic phase to obtain nutrients and colonize the tissue. Some pathogens that have traditionally been classified as necrotrophs exhibit some biotrophic characteristics.

Genomic and transcriptomic analyses of species in the hemibiotrophic *Colletotrichum* genus indicates that they use a wide range of biochemical processes to combat host defenses [27-29]. Although differences were observed between species, hundreds of genes that encode small, secreted proteins were found and proposed to be effector molecule homologues to other biotrophic fungi [27]. Interestingly, species in the genus *Colletotrichum* contain the largest collection of proteases and carbohydrate-active enzymes of any fungus. They also produce secondary metabolites, which are commonly phytotoxic, in both their biotrophic and ne- crotrophic phases [27].

There are a number of species of *Colletotrichum* that infect soybean [30, 31]. *Colletotrichum truncatum* (Schwein.) Andrus & W.D. Moore is one of the species that causes anthracnose of soybean. The fungus can infect the soybean plant at any growth stage, but symptoms are most commonly observed as the plant matures. The disease symptoms are more prevalent in warm and humid conditions, and it causes poor stand and seed quality issues with reduction in yields by up to 20% [32]. Symptoms of anthracnose include irregularly shaped, brown to black blotches, and sunken cankers on stems, petioles, and pods. Infected leaves will be shrunken, rolled, or wilted with necrotic laminar veins [31].

The pathogen is commonly seed-borne but can persist in plant debris as mycelia or acervuli, an asexual fruiting body. In conducive conditions, the acervuli will produce colorless coni- diophores and abundant conidia (Figure 5A). Conidia are frequently dispersed onto plant tissues by splashing rain. Infection begins when conidia germinate on plant tissue, forming a melanized appressorium to directly infect the cuticle layer of the host. Inside the sub-cuticular layer, the appressoria (Figure 5B) form primary infection hyphae, unrecognizable to the host, that grow along the cell walls and branch profusely to colonize the intercellular space [33]. The necrotrophic phase of the pathogen starts when the secondary hyphae emerge from the primary hyphae and quickly penetrate the epidermal and mesophyll cells. The cells collapse causing necrotic lesions. As early as 36 hours after infection, acervuli will develop [32]. These structures overwinter on plant debris until conditions in the spring allow for the cycle to repeat.

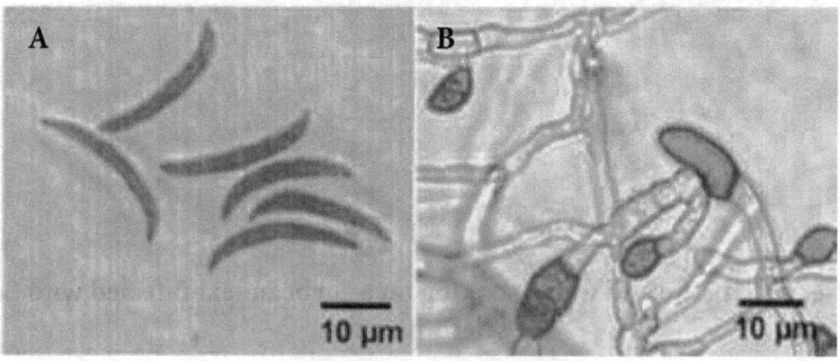

Figure 5. A) *Colletotrichum truncatum* conidia stained with trypan blue and B) appressoria (Yang et al. 2014).

5. Beneficial fungi

The phytobiome is the area surrounding a plant that associates or influences plant growth; it includes microbes, insects, nematodes, and abiotic factors, such as temperature and moisture [34]. A major part of the phytobiome is the rhizosphere. The rhizosphere is the soil surrounding the root tissues, a few millimeters thick, and potentially support 10^{11} microbial cells/g of soil. In grasslands, fungi made up 20-66% of the microbial biomass in the rhizosphere [35]. Some of these microbes can be pathogens, but many of these microbes have beneficial attributes to offer the plant. It has been well known that beneficial microbes in the rhizosphere have a crucial impact on host health, including uptake of nutrients, disease suppression, and host immunity [36]. These include mycorrhizal fungi and fungal endophytes.

Mycorrhizal fungi are one group of fungi that often benefit plants and have a mutualistic, biotrophic relationship. Mycorrhizal fungi are categorized into arbuscular mycorrhizal fungi (AMF), ectomycorrhizal fungi, and orchid mycorrhizal fungi. Many of these fungal species are credited with helping plants first move onto land [37]. With over 200 species, AMF are ubiquitous and survive in a broad range of environments and form mutualistic associations with approximately 80% of all land plants. AMF are found in the rhizosphere and have a positive impact on the host by moving nutrients and water from outside the rhizosphere to give to the host in return for carbon. In order for these nutrients to move efficiently, these fungi infect and colonize the cortex of plants. This sets off innate defense responses, and mycorrhizal fungi have a sophisticated repertoire of mechanisms to maintain a mutualistic relationship with the host.

Soybean plants colonized by AMF have shown greater drought tolerance and an increase capacity to uptake nutrients [38-40]. All AMF have a broad host range, but there is some host preference among crops. *Funneliformis mosseae* (T.H. Nicolson & Gerd.) C. Walker & A. Schüßler and *Glomus* spp. Tul. & C. Tul. are AMF commonly associated with soybean [41, 42]. Much of the basic research of AMF with legumes has been on *M. truncatula* and not soy- beans, although the findings may be applicable to soybean plants as well.

AMF interact with their host well before any physical contact occurs. Since AMF are obligate biotrophs, presymbiotic signaling between the plant and the fungus is required for spore germination [43]. The first step to initiate this symbiotic relationship originates from the plant when grown under low orthophosphate conditions. The plant releases strigolactones through root exudates that stimulate AMF spore germination [43-45]. Flavonoids from the plant have also been associated with AMF spore germination, but their role in pre-symbiosis is not known. The fungus recognizes strigolactones and/or other potential signals that activate mobilization of triacylglycerides and glycogen within the spores to produce germ tubes and extra radical hyphae. This carbon storage can maintain growth for a couple days while the fungus is finding and establishing a relationship with the plant. In absence of plant signals, a germinated spore is able to retract fungal cytoplasm for a later attempt [43].

Reciprocating fungal diffusible signals following host recognition stimulate a counter response from the plant. Two fungal diffusible signals have been identified as short- and long-chain chitin oligomers [46, 47]. These signals will stimulate expression of the early nodulation gene, ENOD11, and activate the common symbiosis signaling pathway [43, 45]. This pathway is comprised of cation channels, a leucine-rich repeat receptor kinase, and nuclear porins all of which are required to induce calcium spiking along with a calcium/calmodulin-dependent kinase [45]. Lateral root formation is also stimulated by diffusible signals, potentially by a separate pathway.

Once in contact with the root tissues, AMF hyphae will move along the root surface of young, lateral roots to find an ideal infection location. The root surface provides cues to the fungus to adhere to the host and establish what is known as a hyphopodium (Figure 1C) [45, 47]. This was shown when *Gigaspora gigantea* (T.H. Nicolson & Gerd.) Gerd. & Trappe colonized cell wall fragments of a carrot host but did not colonize root fragments of a non-host, common beet [48]. Hyphopodia formation is severely reduced on plant mutants (*required for arbuscular mycorrhiza, ram1* and *ram2*). [45]. RAM1 encodes for a gibberellic acid-insensitive transcription factor upstream of RAM2, a glycerol-3-phosphate acyl transferase, and produces cutin monomers [49]. Cutin monomers trigger appressorium induction in a number of pathogenic fungi indicating it could be a conserved feature among all fungi that colonize plants [45].

Following hyphopodia formation, the plant epidermal cells prepare for colonization by restructuring the cytoplasm (Figure 1C) [47, 50]. The nucleus of the cell moves toward the contact site followed by reorganization of microtubules and actin bundles between the nucleus and the contact site. The nucleus then migrates towards the underlying cell causing more reorganization of microtubules to eventually form a column through the plant cell. This structure is termed the pre-penetration apparatus. Once the pre-penetration apparatus is established, the fungus will penetrate the cell directly and follow through the pre-penetration apparatus to not destroy the integrity of the cell. This compartment is now part of the apoplast [47, 50]. The hyphae will branch out in the apoplast colonizing the cortex of the plant and arbuscules will become established (Figure 1C).

Arbuscules are ephemeral structures that are engulfed by plant cells to deliver nutrients to the host in exchange for carbon. Arbuscules are formed by repeated branching of intercellular hyphae that connects to the plant cell by a periarbuscular membrane [51]. The periarbuscular membrane is the center of nutrient transfer. To form this association, extreme re-programming of the inhabited cortical cell transcriptome is required. Over 500 protein-coding genes are upregulated in mycorrhizal plant cells [51]. Proliferation of endoplasmic reticulum, and Golgi apparatus, increase and deployment of plastids, and enlargement of the nucleus are some actions of the host cell needed to accommodate an arbuscule [51]. Individual arbuscules live up to five days, but are continuously being renewed in other cells so there is still a plethora of benefits to the plant including increased disease resistance, drought tolerance, and nutrient uptake. Roughly 20% of the carbon sequestered from photosynthesis is transferred to the fungus through these arbuscules.

Phosphate is an essential nutrient involved in photosynthesis and energy production. Ortho- phosphate is found in low concentrations in the soil and relatively immobile so it is depleted quickly from the rhizosphere [44, 45]. Using its extra-radial mycelia, AMF can reach beyond the rhizosphere to uptake orthophosphate. Within the extra-radial mycelia, orthophosphate is converted to polyphosphate until it reaches the inter-radial mycelia where it is converted back to orthophosphate and brought to the plant via phosphate transporters.

Though phosphate has been the main focus in AMF research, these fungi also can transfer other nutrients, such as nitrogen. Two nitrogen transporters are located in the extra and inter-radial mycelia of *Rhizophagus irregularis* (Błaszk., Wubet, Renker & Buscot) C. Walker & A. Schüßler [52]. Three ammonium transporter genes were discovered in *Geosiphon pyriformis* (Kütz.) F. Wettst. showing is also taken up in the form of ammonium. A nitrate reductase, two nitrite reductases, a glutamine synthetase (GS), and a putative glutamate synthase (GOGAT) transcripts have all been identified in *R. irregularis*, which points to a potential extension of the GS/GOGAT cycle within the extra-radial mycelia of the fungus [44].

Fungal endophytes are another group of organisms that form a symbiotic relationship with their host plant and colonize all parts of the plant. Endophytes emit beneficial compounds such as phytohormones like auxin and gibberellins and secondary metabolites that have shown antimicrobial activity [53]. Genera found to include endophytes associated with soybean are *Alternaria, Cladsporium, Diaporthe,* and *Epicoccum* [54]. Although their mechanisms for host-microbe interactions are not well-known, endophytic fungi have gained attention in recent years as potential amendments in agriculture because of their positive impacts on plant health by promoting nutrient uptake, water acquisition, and by providing protection from abiotic and biotic stresses [53].

6. Influences on infection and colonization

The fungi portrayed in this chapter show many examples of the complex and effective host mechanisms that fungi have to overcome or even manipulate. Most of the fungal-plant interactions have been studied under highly controlled laboratory environments where most factors can be controlled. In the field, the plant interacts with an array of microbes simultaneous- ly along with abiotic stresses, which can alter interactions among organisms. The success of a pathogen is based on the disease triangle (virulent pathogen, susceptible host, and a condu- cive environment).

The benefits of mycorrhizal associations are influenced by the environment and are consid- ered to have a mutualistic-parasitic continuum [55]. Mycorrhizal fungi do not always benefit the plant, and can negatively impact biomass depending on external factors, including microbial populations, nutrient composition, and soil pH. AMF provide nutrients to their hosts in return for carbon, but in high quality soils, this assistance may not have a substantial benefit to the plant and the fungus may reduce carbon thus reducing biomass as well as directly compete for host resources.

7. Plant defense pathways

How fungi infect plants depends on the type of fungus, the plant, and the association. The association includes how the plant defends itself or whether it allows fungal infection and colonization. In response to fungi, plants have evolved different ways to recognize fungi through a network of different plant processes. Several processes important in plant defense are activated through the salicylic acid (SA) and the jasmonic acid (JA) pathways [56]. The SA pathway uses SA as a signaling molecule that is active in defense against biotrophic pathogens. SA is synthesized through the phenylalanine ammonia-lyase (PAL) and the isochorismate syn- thase (ICS) pathways, and accumulation of SA will induce a systemic immune response throughout the plant termed systemic acquired resistance (SAR). This induces *pathogenesis- related* (*PR*) genes, that encode for enzymes such as chitinases and β-1,3-glucanases that affect fungal cell walls [57]. The JA pathway is SA-independent and is traditionally attributed to immunity against necrotrophic pathogens. JA induces transcripts *DAD1, LOX2, AOS,* and *OPR3* as well as PR genes [58]. JA also stimulates induced systemic resistance, a systemic resistance similar to SAR by defending the plant against a broad range of pathogens [59]. There are also immune response-independent mechanisms that are involved in plant defense. Some patho- gens have evolved to manipulateorsurpassdefense responses; therefore, plants require immune response-independent factors for survival. Four transcription factors in *Arabidopsis* were found to be required for host resistance to necrotrophs [60]. Mutants without one or more of these transcription factors were susceptible to infection without compromising either the SA- or JA- defense pathways indicating there are other biological processes involved in host defense.

AMF have a sophisticated relationship with the host defense pathways to maintain a symbiot- ic relationship. Similar to most biotrophs, AMF stimulate SAR by the SA-dependent defense pathway and activate defense compound production such as ROS [61]. SA-induced defense is a negative regulator of AMF symbiosis and is quickly repressed by the fungus by Myc factors and effectors. A superoxide dismutase has been characterized in *R. irregularis* and *Gigaspora margarita* W.N. Becker & I.R. Hall to reduce the amount of ROS [44]. In *R. intraradices* (N.C. Schenck & G.S. Sm.) C. Walker & A. Schüßler, the SP7 effector has shown defense-suppres- sive activity and interacts with the ethylene-responsive factor ERF19 [62]. JA and ET path- ways are upregulated in well-established symbiosis and have a positive impact on colonization of pathogens related to these pathways, such as necrotrophs and chewing herbivores.

8. Summary

Associations of fungi and plants are a continuum from beneficial to mutualistic to pathogen- ic. Based on how fungi obtain nutrients from plants, they can be classified as biotrophs, necrotrophs, or hemibiotrophs. Biotrophic fungi, such as *P. pachyrhizi* and arbuscular mycorrhi- zal fungi, grow only on living plant tissue, extracting nutrients from living plant cells. On the other hand, necrotrophic fungi, such as *S. sclerotiorum*, acquire nutrients for growth and reproduction from dead plant tissue. Some of these necrotrophic fungi also have a biotrophic phase in the life cycle, making them hemibiotrophs, such as *C. truncatum*.

Plants have developed a complex network of biosynthetic pathways, some of which are known to respond to fungal infection and colonization. In turn, fungi have developed sophisticated infection and colonization mechanisms to interact with the host. Advances in technology have allowed for a better perspective on this association in terms of understanding the specific protein- protein interactions and genetic sequencing to find genes controlling these processes. These genes can be directly inserted into plants through genetic engineering. As an example, non– host-linked genes in *A. thaliana* have been transferred to soybean to enhance resistance to soybean rust [63]. In the future, the exploitation of interspecies non–host-linked genes could be used to genetically engineer crops to control diseases or enhance beneficial fungal associa- tions with plants.

References

1. Parfrey, LW, Lahr, DJ, Knoll, AH, Katz, LA. Estimating the timing of early eukaryotic diversification with multigene molecular clocks. Proceedings of the National Academy of Sciences. 2011; 108: 13624-13629.
2. Redecker, D, Kodner, R, Graham, LE. Glomalean fungi from the Ordovician. Science. 2000; 289: 1920-1921.
3. Hartman, GL, Rupe, JC, Sikora, EJ, Domier, LL, Steffey, KL, Davis, JA. Compendium of Soybean Diseases and Pests, fifth edition 2015.
4. Mendgen, K, Hahn, M. Plant infection and the establishment of fungal biotrophy. Trends in Plant Science. 2002; 7: 352-356.
5. Vittal, R, Paul, C, Hill, C, Hartman, G. Characterization and quantification of fungal

colonization of *Phakopsora pachyrhizi* in soybean genotypes. Phytopathology. 2014; 104: 86-94.

6. Koch, E, Ebrahim-Nesbat, F, Hoppe, H. Structural studies on the development of soybean rust (*Phakopsora pachyrhizi* syd.) in susceptible soybean leaves. Phytopatholo- gy Z. 1983;106:302-320. 1983.
7. Chang, H-X, Miller, LA, Hartman, GL. Melanin-independent accumulation of turgor pressure in appressoria of *Phakopsora pachyrhizi* . Phytopathology. 2014; 104: 977-984.
8. McLean, R. Histological studies of resistance to soybean rust, *Phakopsora pachyrhizi* syd. Crop and Pasture Science. 1979; 30: 77-84.
9. Kelly, H, Dufault, NS, Walker, D, Isard, S, Schneider, RW, Giesler, L, Wright, DL, Marois, JJ, Hartman, G. From select agent to an established pathogen: The response to *Phakopsora pachyrhizi* in North america. Phytopathology. 2015; 105:905-916.
10. Miles, M, Bonde, M, Nester, S, Berner, D, Frederick, R, Hartman, G. Characterizing resistance to *Phakopsora pachyrhizi* in soybean. Plant Disease. 2011; 95: 577-581.
11. Schneider, KT, Van DE Mortel, M, Bancroft, TJ, Braun, E, Nettleton, D, Nelson, RT, Frederick, RD, Baum, TJ, Graham, MA, Whitham, SA. Biphasic gene expression changes elicited by *Phakopsora pachyrhizi* in soybean correlate with fungal penetration and haustoria formation. Plant Physiology. 2011; 157: 355-371.
12. Choi, J, Alkharouf, N, Schneider, K, Matthews, B, Frederick, R. Expression patterns in soybean resistant to *Phakopsora pachyrhizi* reveal the importance of peroxidases and lipoxygenases. Functional & Integrative Genomics. 2008; 8: 341-359.
13. Slaminko, T, Miles, M, Frederick, R, Bonde, M, Hartman, G. New legume hosts of *Phakopsora pachyrhizi* based on greenhouse evaluations. Plant Disease. 2008; 92: 767-771.
14. Loehrer, M, Langenbach, C, Goellner, K, Conrath, U, Schaffrath, U. Characterization of nonhost resistance of Arabidopsis to the asian soybean rust. Molecular Plant-Microbe Interactions. 2008; 21: 1421-1430.
15. Hoefle, C, Loehrer, M, Schaffrath, U, Frank, M, Schultheiss, H, Hückelhoven, R. Transgenic suppression of cell death limits penetration success of the soybean rust fungus *Phakopsora pachyrhizi* into epidermal cells of barley. Phytopathology. 2009; 99: 220-226.
16. Ishiga, Y, Uppalapati, SR, Gill, US, Huhman, D, Tang, Y, Mysore, KS. Transcriptomic and metabolomic analyses identify a role for chlorophyll catabolism and phytoalexin during Medicago nonhost resistance against asian soybean rust. Scientific reports. 2015; 5:1-17.
17. Laluk, K, Mengiste, T. Necrotroph attacks on plants: Wanton destruction or covert extortion? The Arabidopsis book / American Society of Plant Biologists. 2010; 8:1-34.
18. Radwan, O, Rouhana, LV, Hartman, GL, Korban, SS. Genetic Mechanisms of Host–Pathogen Interactions for Charcoal Rot in Soybean. Plant molecular biology reporter. 2014; 32: 617-629.
19. Wolpert, TJ, Dunkle, LD, Ciuffetti, LM. Host-selective toxins and avirulence determi-

nants: What's in a name?*. Annual Review of Phytopathology. 2002; 40: 251-285.

20. Colmenares, AJ, Aleu, J, Duran-Patron, R, Collado, IG, Hernandez-Galan, R. The Pu- tative Role of Botrydial and Related Metabolites in the Infection Mechanism of *Botry- tis cinerea*. Journal of Chemical Ecology. 2002; 28: 997-1005.

21. Bolton, MD, Thomma, B, Nelson, BD. *Sclerotinia sclerotiorum* (lib.) de bary: Biology and molecular traits of a cosmopolitan pathogen. Molecular Plant Pathology. 2006; 7: 1-16.

22. Garg, H, Li, H, Sivasithamparam, K, Kuo, J, Barbetti, MJ. The infection processes of *Sclerotinia sclerotiorum* in cotyledon tissue of a resistant and a susceptible genotype of *Brassica napus*. Annals of Botany. 2010; 106: 897-908.

23. Calla, B, Vuong, T, Radwan, O, Hartman, GL, Clough, SJ. Gene expression profiling soybean stem tissue early response to *Sclerotinia sclerotiorum* and in silico mapping in relation to resistance markers. Plant Genome. 2009; 2: 149-166.

24. Amselem, J, Cuomo, CA, van Kan, JAL, Viaud, M, Benito, EP, Couloux, A, Coutinho, PM, de Vries, RP, Dyer, PS, Fillinger, S, Fournier, E, Gout, L, Hahn, M, Kohn, LM, Lapalu, N, Plummer, KM, Pradier, JM, Quevillon, E, Sharon, A, Simon, A, ten Have, A, Tudzynski, B, Tudzynski, P, Wincker, P, Andrew, M, Anthouard, V, Beever, RE, Beffa, R, Benoit, I, Bouzid, O, Brault, B, Chen, ZH, Choquer, M, Collemare, J, Cotton, P, Danchin, EG, Da Silva, C, Gautier, A, Giraud, C, Giraud, T, Gonzalez, C, Grosse- tete, S, Guldener, U, Henrissat, B, Howlett, BJ, Kodira, C, Kretschmer, M, Lappar- tient, A, Leroch, M, Levis, C, Mauceli, E, Neuveglise, C, Oeser, B, Pearson, M, Poulain, J, Poussereau, N, Quesneville, H, Rascle, C, Schumacher, J, Segurens, B, Sex-ton, A, Silva, E, Sirven, C, Soanes, DM, Talbot, NJ, Templeton, M, Yandava, C, Yard- en, O, Zeng, QD, Rollins, JA, Lebrun, MH, Dickman, M. Genomic analysis of the necrotrophic fungal pathogens *Sclerotinia sclerotiorum* and *Botrytis cinerea*. Plos Genet- ics. 2011; 7. 1-27.

25. Rollins, JA, Cuomo, CA, Dickman, MB, Kohn, LM. Genomics of *Sclerotinia Sclerotio- rum*. In: Genomics of Plant-Associated Fungi and Oomycetes: Dicot pathogens, Springer; 2014. 1-17.

26. Kabbage, M, Yarden, O, Dickman, MB. Pathogenic attributes of *Sclerotinia sclerotio- rum*: Switching from a biotrophic to necrotrophic lifestyle. Plant Science. 2015; 233: 53-60.

27. McDowell, JM. Genomic and transcriptomic insights into lifestyle transitions of a hemi-biotrophic fungal pathogen. New Phytologist. 2013; 197: 1032-1034.

28. Gan, P, Ikeda, K, Irieda, H, Narusaka, M, O'Connell, RJ, Narusaka, Y, Takano, Y, Ku- bo, Y, Shirasu, K. Comparative genomic and transcriptomic analyses reveal the hemi- biotrophic stage shift of *Colletotrichum* fungi. New Phytologist. 2013; 197: 1236-1249.

29. O'Connell, RJ, Thon, MR, Hacquard, S, Amyotte, SG, Kleemann, J, Torres, MF, Damm, U, Buiate, EA, Epstein, L, Alkan, N, Altmuller, J, Alvarado-Balderrama, L, Bauser, CA, Becker, C, Birren, BW, Chen, ZH, Choi, J, Crouch, JA, Duvick, JP, Farm- an, MA, Gan, P, Heiman, D, Henrissat, B, Howard, RJ, Kabbage, M, Koch, C, Krach-

er, B, Kubo, Y, Law, AD, Lebrun, MH, Lee, YH, Miyara, I, Moore, N, Neumann, U, Nordstrom, K, Panaccione, DG, Panstruga, R, Place, M, Proctor, RH, Prusky, D, Rech, G, Reinhardt, R, Rollins, JA, Rounsley, S, Schardl, CL, Schwartz, DC, Shenoy, N, Shirasu, K, Sikhakolli, UR, Stuber, K, Sukno, SA, Sweigard, JA, Takano, Y, Takahara, H, Trail, F, van der Does, HC, Voll, LM, Will, I, Young, S, Zeng, QD, Zhang, JZ, Zhou, SG, Dickman, MB, Schulze-Lefert, P, van Themaat, EVL, Ma, LJ, Vaillancourt, LJ. Lifestyle transitions in plant pathogenic *Colletotrichum* fungi deciphered by genome and transcriptome analyses. Nature Genetics. 2012; 44: 1060-1065.

30. Yang, HC, Haudenshield, JS, Hartman, GL. *Colletotrichum incanum* sp nov., a curved-conidial species causing soybean anthracnose in USA. Mycologia. 2014; 106: 32-42.

31. Yang, HC, Haudenshield, JS, Hartman, GL. Multiplex real-time pcr detection and dif- ferentiation of *Colletotrichum* species infecting soybean. Plant Disease. 2015; 99: 1559-1568.

32. Manandhar, JB, Kunwar, IK, Singh, T, Hartman, GL, Sinclair, JB. Penetration and infection of soybean leaf tissues by *Colletotrichum truncatum* and *Glomerella glycines*. Phytopathology. 1985; 75: 704-708.

33. Ranathunge, NP, Mongkolporn, O, Ford, R, Taylor, PWJ. *Colletotrichum truncatum* pa- thosystem on *Capsicum* spp: Infection, colonization and defence mechanisms. Aus- tralasian Plant Pathology. 2012; 41: 463-473.

34. Beattie, G., and J. E. Leach. "The Phytobiomes Initiative." Phytopathology: 2014; 104:166-166.

35. Joergensen, RG. Ergosterol and microbial biomass in the rhizosphere of grassland soils. Soil Biology & Biochemistry. 2000; 32: 647-652.

36. Berendsen, RL, Pieterse, CMJ, Bakker, PAHM. The rhizosphere microbiome and plant health. Trends in Plant Science. 2012; 17: 478-486.

37. Akhtar, MS, Siddiqui, ZA, Wiemken, A. Arbuscular mycorrhizal fungi and rhizobium to control plant fungal diseases. New York, Springer; 2011. 263-292.

38. Liu, ZL, Li, YJ, Ma, LN, Wei, HC, Zhang, JF, He, XY, Tian, CJ. Coordinated regulation of arbuscular mycorrhizal fungi and soybean mapk pathway genes improved mycorrhizal soybean drought tolerance. Molecular Plant-Microbe Interactions. 2015; 28: 408-419.

39. Meng, LB, Zhang, AY, Wang, F, Han, XG, Wang, DJ, Li, SM. Arbuscular mycorrhizal fungi and rhizobium facilitate nitrogen uptake and transfer in soybean/maize intercropping system. Frontiers in Plant Science. 2015; 6:1-10.

40. Abdel-Fattah, GM, Asrar, AA, Al-Amri, SM, Abdel-Salam, EM. Influence of arbuscular mycorrhiza and phosphorus fertilization on the gas exchange, growth and phosphatase activity of soybean (glycine max l.) plants. Photosynthetica. 2014; 52: 581-588.

41. Jie, WG, Liu, XR, Cai, BY. Diversity of rhizosphere soil arbuscular mycorrhizal fungi in various soybean cultivars under different continuous cropping regimes. Plos One. 2013; 8:1-9.

42. Kojima, T, Oka, N, Karasawa, T, Okazaki, K, Ando, S, Takebe, M. Community of ar-

buscular mycorrhizal fungi in soybean roots after cultivation with different cropping systems. Jarq-Japan Agricultural Research Quarterly. 2014; 48: 279-290.

43. Schmitz, AM, Harrison, MJ. Signaling events during initiation of arbuscular mycorrhizal symbiosis. Journal of Integrative Plant Biology. 2014; 56: 250-261.

44. Malbreil, M, Tisserant, E, Martin, F, Roux, C. Genomics of arbuscular mycorrhizal fungi: Out of the shadows. In: Fungi. F. M. Martin; 2014. 70: 259-290.

45. Gutjahr, C, Parniske, M. Cell and developmental biology of arbuscular mycorrhiza symbiosis. In: Annual Review of Cell and Developmental Biology. R. Schekman; 2013. 29: 593-617.

46. Genre, A, Chabaud, M, Balzergue, C, Puech-Pages, V, Novero, M, Rey, T, Fournier, J, Rochange, S, Becard, G, Bonfante, P, Barker, DG. Short-chain chitin oligomers from arbuscular mycorrhizal fungi trigger nuclear ca^{2+} spiking in *Medicago truncatula* roots and their production is enhanced by strigolactone. New Phytologist. 2013; 198: 179-189.

47. Maillet, F, Poinsot, V, Andre, O, Puech-Pages, V, Haouy, A, Gueunier, M, Cromer, L, Giraudet, D, Formey, D, Niebel, A, Martinez, EA, Driguez, H, Becard, G, Denarie, J. Fungal lipochitooligosaccharide symbiotic signals in arbuscular mycorrhiza. Nature. 2011; 469: 58-63.

48. Nagahashi, G, Douds Jr, DD. Appressorium formation by am fungi on isolated cell walls of carrot roots. New Phytologist. 1997: 299-304.

49. Wang, E, Schornack, S, Marsh, JF, Gobbato, E, Schwessinger, B, Eastmond, P, Schultze, M, Kamoun, S, Oldroyd, GE. A common signaling process that promotes mycorrhizal and oomycete colonization of plants. Current Biology. 2012; 22: 2242-2246.

50. Smith, SE, Read, D. Mycorrhizal symbiosis : third edition. San Diego, Elsevier Academic Press Inc; 2008.

51. Bonfante, P, Genre, A. Mechanisms underlying beneficial plant-fungus interactions in mycorrhizal symbiosis. Nature Communications. 2010; 1:1-48.

52. Madala, NE, Steenkamp, PA, Piater, LA, Dubery, IA. Metabolomic analysis of isonitrosoacetophenone-induced perturbations in phenolic metabolism of *Nicotiana taba- cum* cells. Phytochemistry. 2013; 94: 82-90.

53. Rai, M, Rathod, D, Agarkar, G, Dar, M, Brestic, M, Pastore, GM, Marostica, MR. Fun- gal growth promotor endophytes: A pragmatic approach towards sustainable food and agriculture. Symbiosis. 2014; 62: 63-79.

54. Impullitti, AE, Malvick, DK. Fungal endophyte diversity in soybean. Journal of Applied Microbiology. 2013; 114: 1500-1506.

55. Smith, FA, Smith, SE. How useful is the mutualism-parasitism continuum of arbuscular mycorrhizal functioning? Plant and Soil. 2013; 363: 7-18.

56. Kachroo, P, Kachroo, A. The roles of salicylic acid and jasmonic acid in plant immunity. Molecular Plant Immunity. 2012: 55-79.

57. An, C, Mou, Z. Salicylic acid and defense responses in plants. In: Phytohormones:

A Window to Metabolism, Signaling and Biotechnological Applications, Springer; 2014. 191-219.

58. Wasternack, C. Jasmonates in plant growth and stress responses. In: Phytohormones: A Window to Metabolism, Signaling and Biotechnological Applications, Springer; 2014. 221-263.

59. Pieterse, CM, Zamioudis, C, Does, DV, Van Wees, S. Signalling networks involved in induced resistance. Induced Resistance for Plant Defense: A Sustainable Approach to Crop Protection. 2014: 58-80.

60. Dobón, A, Canet, JV, García-Andrade, J, Angulo, C, Neumetzler, L, Persson, S, Vera, P. Novel disease susceptibility factors for fungal necrotrophic pathogens in arabidopsis. Plos Pathogens. 2015; 11: e1004800-e1004800.

61. Jung, SC, Martinez-Medina, A, Lopez-Raez, JA, Pozo, MJ. Mycorrhiza-induced resistance and priming of plant defenses. Journal of Chemical Ecology. 2012; 38: 651-664.

62. Zamioudis, C, Pieterse, CMJ. Modulation of host immunity by beneficial microbes. Molecular Plant-Microbe Interactions. 2012; 25: 139-150.

63. Langenbach, C, Schultheiss, H, Rosendahl, M, Tresch, N, Conrath, U, Goellner, K. In- terspecies gene transfer provides soybean resistance to a fungal pathogen. Plant Bio- technology Journal. 2016; 14:699-708.

Chapter-3

Stenocarpella maydis and Sporisorium reilianum: Two Pathogenic Fungi of Maize

1. Introduction

The crop diseases caused by fungi represent a major obstacle to agriculture worldwide. Corn (*Zea mays* L.) is a crop with a high level of consumption, which is affected by various diseases. *Stenocarpella maydis* and *Sporisorium reilanum* are fungi that generate white rot in corn cob and head smut in maize, respectively, both of which are diseases distributed worldwide causing numerous economic losses [1–4]. *S. reilianum* is a phytopathogen belonging to the order Ustilaginales, which infect a large number of monocotyledonous and dicotyledonous plants. The most outstanding feature of the disease is the presence of carbonous masses of black coloration in the corn cob and maize tassel. T his causes excessive deformation and over development, which has been a serious problem since the early 1970s in countries like the United States, Australia, China, South Africa, France, and Mexico [3]. *S. maydis* causes diseases, and this fungus has a worldwide distribution. The fungus remains latent as mycelium, pycnidia, and spores in crop residues or seeds. The cycle starts from sexual or asexual spores that over -winter on cereal or traces of stubble. The spores are carried by abiotic (wind, rain drops) and biotic (insects, birds) agents towards the maize tassel, where they find the main entrances to the plant; the stigma and damaged, developing grains. The disease is favored when the weather is wet after flowering and when the atmosphere is cool and humid during the grain- filling stage [1]. Currently, control of both fungi has been conducted mainly with the use of resistant hybrids; however, the genotypes that were resistant to the disease in one year may be susceptible in the following season. These fungi produce severe damage to maize cultivation because of their ability to degrade the cell wall components. They do this by excreting enzymes, allowing the infection of, and colonization in, the host plant. Therefore, consideration has been given to the possibility of studying the production and the character- istics of these fungal enzymes, which include xylanases, cellulases, proteases, etc., as well as their associated potential biotechnological applications.

2. Overview of *Sporisorium reilianum* and *Stenocarpella maydis*

2.1. Sporisorium reilianum

2.1.1. General characteristics

S. reilianum is a pathogenic basidiomycete, both biotrophic and dimorphic, and is the causal agent of head smut in maize [5–7]. It belongs to the Ustilaginaceae family and was first described as *Ustilago reiliana* (Kühn) and then renamed as *Sphacelotheca reiliana* (Kühn). Studies based on its genetic characteristics allowed it to be placed in the *Sporisorium* genus with two subspecies: *S. reilianum* f. sp. *reilianum* and *S. reilianum* f. sp. *zeae*, affecting sorghum and maize respectively. However, both varieties are able to infect and invade both hosts. *S. reilianum* f. sp. *reilianum*, is highly virulent in sorghum, but does not produce spores on maize, while *S. reilianum* f. sp. *zeae* causes no disease in sorghum with the only recognized symptoms observed being the presence of phytoalexins. Transcriptome analysis of maize leaves colonized by both pathogens showed that most genes are induced with *S. reilianum* f. sp. *zeae* compared with that of *S. reilianum* f. sp. *reilianum*, showing that host specificity is determined by different mech- anisms in sorghum and maize [8–9]. This fungus is an inhabitant of the soil where it can survive up to 10 years in the form of a teliospore: a structure generated by fragmentation of the mycelium in plant tissues either on the tassels or in the corn cob. These are semispherical, echinulate, yellowish brown in color but can range from pale to dark red or black (Figure 1). They can be dispersed by rain, wind, wildlife, agricultural machinery, or human beings [5, 10].

The life cycle begins when a dikaryotic young teliospore, suffers karyogamy, giving rise to a spore mature diploid uninucleate. When optimum temperature and humidity conditions are presented, they germinate producing a structure called promycelium, where the nucleus divides by meiosis and the resulting four nuclei pass to lateral cells to form four haploid basidiospores of different sexual compatibility. These can remain in saprophytic manner with their division being by gemmation. I t is said that at this point the fungus is in its yeast phase (Figure 2).

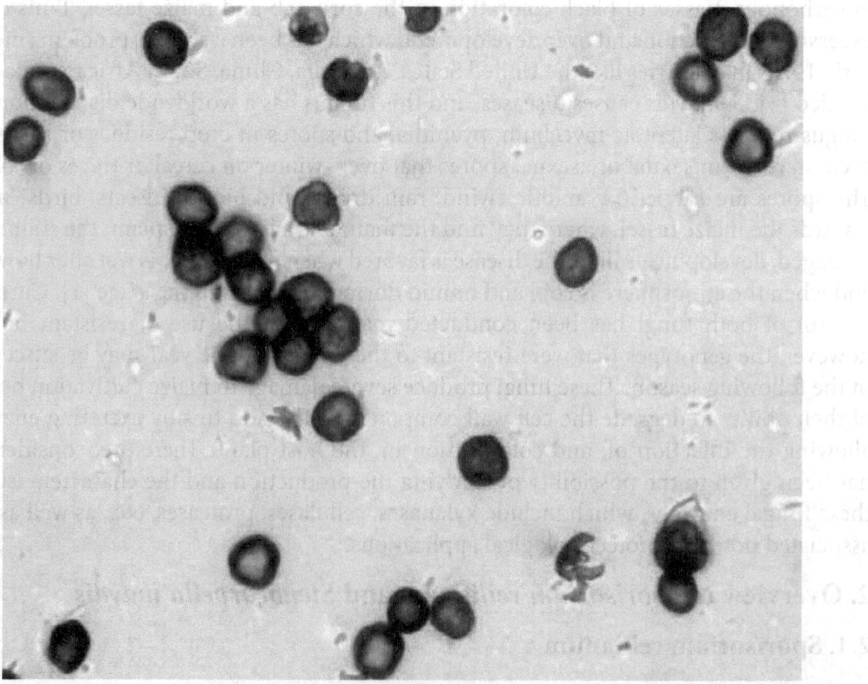

Figure 1. Teliospores of S. reilianum. Observation made with a phase contrast microscope at 40X.

When two yeasts with different sexual compatibility (a and b) produce and recognize phero- mones, they may come into contact with the young tissues of a plant, forming a complemen- tation tube which allows cell fusion (somatogamy), giving rise to an infective stage which is constituted by a septate dikaryotic mycelium. The formation of an appressorium is crucial for penetration, where the production of lytic enzymes and the mechanical processes of pressure, probably play an important role. In this case the fungus locally degrades the cell wall of the epidermis, permitting penetration and a systemic invasion mainly affecting the undifferenti- ated reproductive organs, either male or female, of the plant, where the production of teliospores at the time of flowering, are manifested as carbonaceous masses of black coloration on the ears and corn cobs, forming what is commonly known as sori or galls (Figure 3). These structures are bare, unlike common smut caused by *Ustilago maydis* where they are covered by a white membranous tissue with traces of the vascular system of the plant. The stages of the infective diploid and the saprophytic haploid can be maintained in the laboratory, where reproduction takes place through gemmation. Factors favoring the transition from yeast to mycelium have not been fully described, but may be linked to the temperature, humidity, and pH of the medium [2, 4, 6, 7, 11–14].

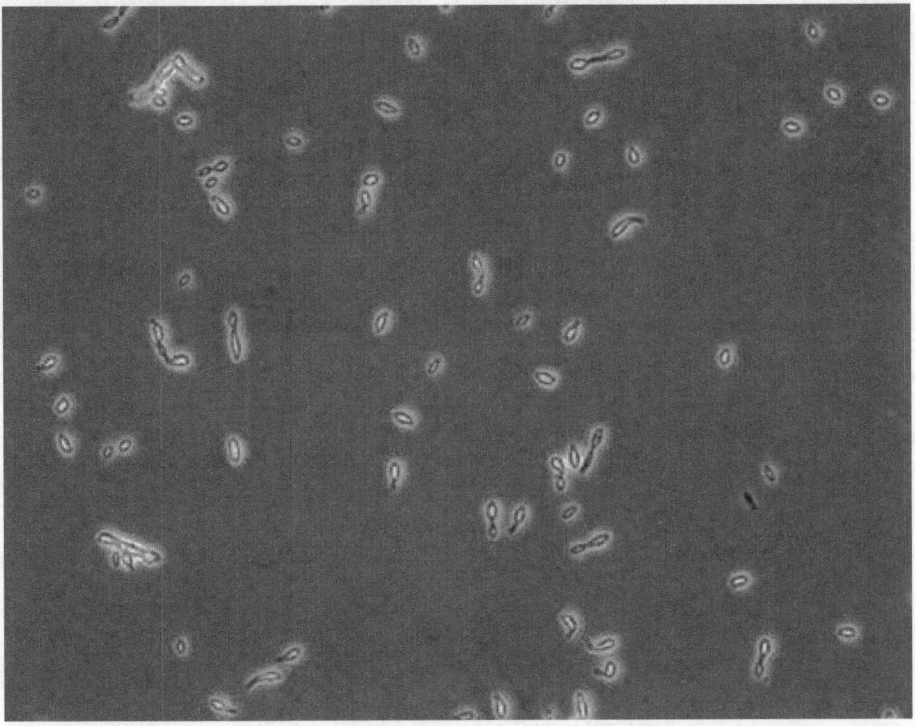

Figure 2. *S. reilianum* **in its yeast phase. Observation made with a light microscope at 40X.**

The complete genome sequence of *S. reilianum* has already been reported and is deposited in the database at the Munich Information Center for Protein Sequences (MIPS) at the following address; *http://mips.helmholtz-muenchen.de/genre/proj/sporisorium/*, which facilitates research in order to understand the mechanisms that occur during the interaction of the fungus with the plant [3].

2.1.2. Characteristics of the disease

The disease caused by *S. reilianum* called head smut are manifested in the flowering stage due to the presence of a carbonaceous mass of teliospores invading male inflorescences in the tassels replacing pollen formation. The same effect is observed quite frequently in the corn cobs, resulting in the appearance of black soil (Figure 4) [11].

It can be seen that anthocyanin accumulates in stalks, together with the presence of chlorotic spots in the leaves. During colonization an increase of 30% of the total content of auxins in the inflorescences, and a significant accumulation of reactive oxygen species, occurs [4, 11, 14].

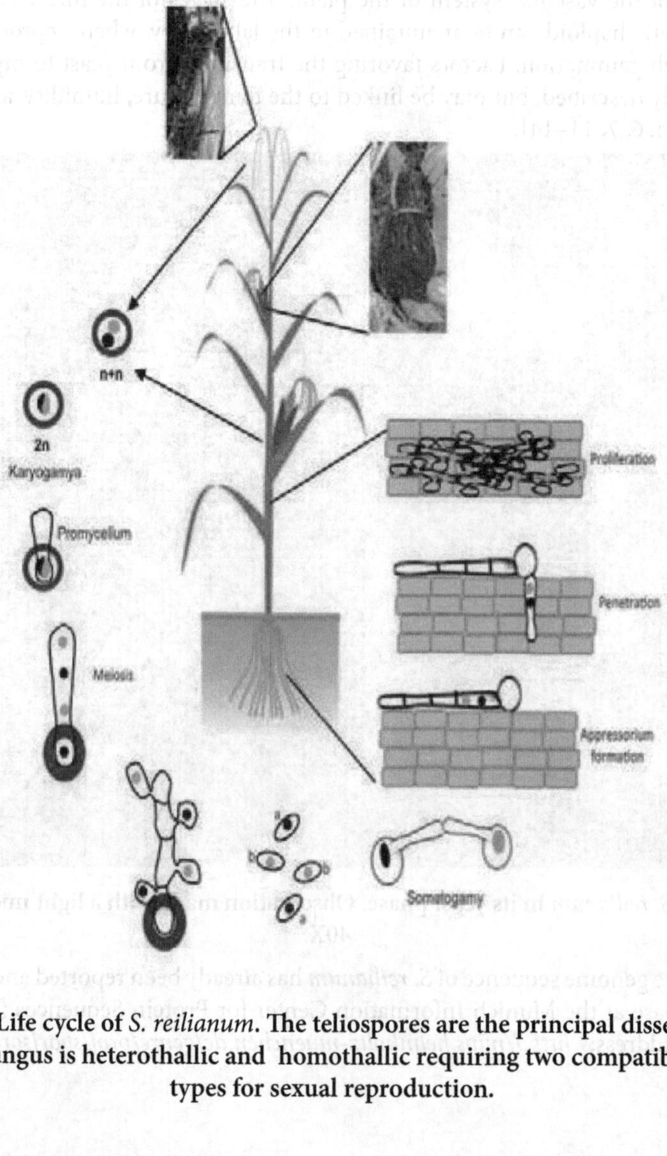

Figure 3. Life cycle of *S. reilianum*. The teliospores are the principal dissemination source. Fungus is heterothallic and homothallic requiring two compatible mating types for sexual reproduction.

The development of the disease is favored when soil moisture is 15–25%, at a temperature of 23–30 °C, with low water potential – the latter has an effect on the transition of basidiospores to hyphae facilitating the fusion of compatible strains, leading to increased disease severity. N utritional aspects are also important: nitrogen deficiency increases infection, with a lower incidence rate being identified in clay soils than sandy soils [3, 10, 11]. H ead smut is not considered as devastating a disease, but still causes severe losses in the crop yields of maize [15–18].

The report for the first specimen was made in 1875 by Kühn, who received the original strain found in Egypt by Dr. Reil in 1868 [19], however, the disease now has a worldwide distribution, especially where maize crops are extensively practiced, such as Europe, North, Central, and South America, Africa, Asia, Australia, New Zealand, western India, and Palestine, among other countries [7, 20–23].

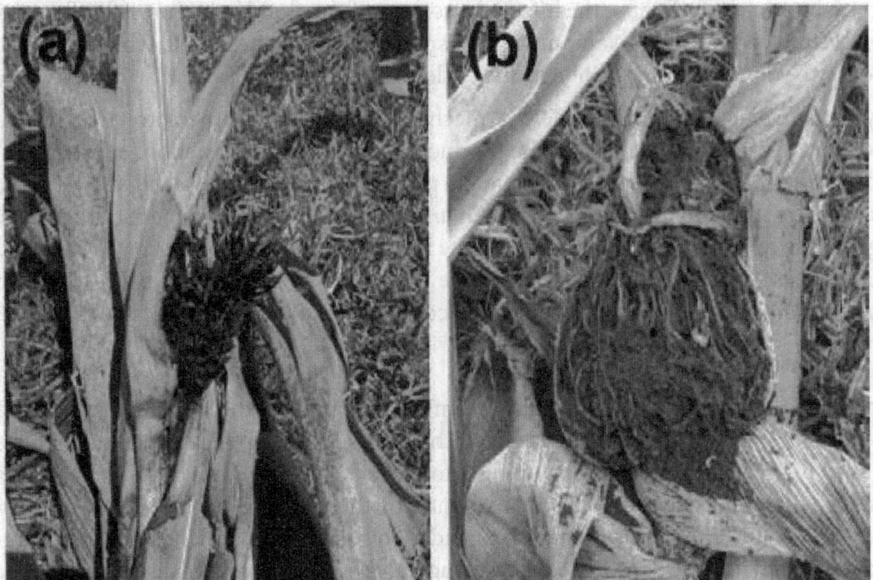

Figure 4. Head smut of maize produced by S. reilianum f. sp. zeae. Presence of a carbonaceous mass of teliospores in the tassel (a) and cob (b).

2.1.3. Pathogenicity factors

During infection and colonization of *S. reilianum* in the floral tissue of maize, one can identify different interactions, which are described as follows: 1) hyphae and cells of host may be associated and alive; 2) hyphae collapse and cells of host remain alive; 3) hyphae and cells of host collapse; and 4) hyphae is viable and colonizes all cells of the host causing tissue death [2]. This plant pathogen has a high compatibility with its host in order to survive until flowering. During growth within the tissue of maize, hyphae are surrounded by a matrix that allows an area of exchange between the plant and fungus, which is separated from the plasmatic membrane. During this interaction, structures like vesicles are observed that could have the function of endocytosis, which may carry virulence effectors to help with compatibility with the host [24].

The growth of filamentous hyphae in epidermal cells of root, apices, and young tassels of maize, show that the hyphae are mainly in the intercellular spaces and between adjacent epidermal cells, exhibiting no damage to the cell wall of the host, although hyphae are attached to the host. For this plant pathogen, considered biotrophic, intracellular growth is a useful strategy to damage and avoid the response of plant defenses [25].

S. reilianum infection in maize induces a loss of apical dominance showing two modifications in the inflorescences, one of which is the loss of identity of, and the appearance of, phyllodes. These results suggest that the fungus modulates the floral architecture of maize whilst floral genetic regulation could be a secondary consequence of increased reactive oxygen species (ROS) or high levels of auxin, or could also be due to the fact that the fungus regulates floral genetic expression by the secretion of some proteins [4]. Symptoms caused during infection are observed after the floral transition where there are white sori that contain the teliospores, which are formed in the younger part of the panicle, and cause the infected tassels not form floral branches. After flowering, typical smut contains a high amount of teliospores in the infected plant, which allows it to colonize new plants and continue its life cycle [25].

Phytopathogenic fungi have different mechanisms that allow them to penetrate and colonize their hosts. One involves the production, and synergistic action, of extracellular hydrolytic enzymes that degrade the different polymers constituting the cell wall of the plant tissues [26]. The genome sequence of *S. reilianum* shows few genes encoding enzymes that degrade the cell wall [27]. Until now only the hydrolytic activities of aspartyl protease and beta-xylanase, produced in different culture media, have been reported. These enzymes could present an important role during the colonization of the host infection [28, 29].

2.1.4. Strategies for disease control

Because *S. reilianum* infects during germination and in the early stages of plant development, the main strategy to control the disease is the application of fungicides to the seed to prevent the pathogen coming into contact with the host. However, it has been observed that some chemicals can retard plant growth and others reduce seed germination [15, 16, 18, 30]. The application of fungicides to foliar structures has not been found to control the disease [31]. Some chemical agents used to control the disease are: Benlate (benomyl) and carboxin + thiram [32]. Furthermore, fungicides have been used to inhibit the synthesis of ergosterol, among which are triazole and imidazole. Lately, has emerged that azoxystrobin and strobilurin present protection via soil treatment [15, 16, 32-34].

Genetic resistance as an alternative disease control method may be more feasible and eco- nomical so development is underway on tolerant maize hybrids with high yields [21]. It has been observed that the use of resistant hybrids to disease in one year may make the next crop susceptible. Crop rotation for legumes, care and cleaning of agricultural machinery, and humidity control, can help reduce the incidence of the disease [35–37].

In recent years, scientists have been looking for new forms of control that also need to be environmentally friendly. One such case is the use of biological controls which represent an alternative for the management of the disease, reducing the use of chemical fungicides [38, 39]. In this respect Mercado-Flores et al. in 2014 applied a native strain of *Bacillus subtilis* to a maize producing area in the Mezquital Valley in the central part of Mexico. It was found that the biological treatment significantly reduced the incidence percentage of smut while increas- ing maize productivity [37].

2.2. Stenocarpella maydis

2.2.1 General characteristics

W hite rot of stalk and corn cob is a disease caused by the ascomycete *S. maydis*, one of the most destructive worldwide, especially during wet seasons. Symptoms manifest many weeks after infection, affecting roots, stems, and corn cobs, where a white cottony fungal growth is observed in the presence of pycnidia, and the marrow stem is discoloured and disintegrated leaving only the vascular bundles intact – the internodes showing a dark brown coloration. I n this case the plant is weakened and easily broken by rain and strong winds. Infected corn grains have less glare and have a dull brown or slightly gray coloration [1, 40, 41].

Natural infection of *S. maydis* on the stem and shank is greater between one to three weeks after pollination, in the presence of rain and temperatures ranging 28–30 °C. Periods of drought before flowering increase the crop's susceptibility to the disease. This occurs mainly in cold regions because conidia lose viability at high temperatures and with exposure to sunlight [1, 40, 42].

S. maydis survives throughout the year between crop residues as pycnidia, which contain the conidia or spores of the fungus (Figure 5). During the wet season, these structures are released and propagated, by splashing rain drops, to the female inflorescences, being deposited around the shank of the corn cob. From there they germinate and penetrate, invading the plant and continuing their life cycle (Figure 6) [43, 44].

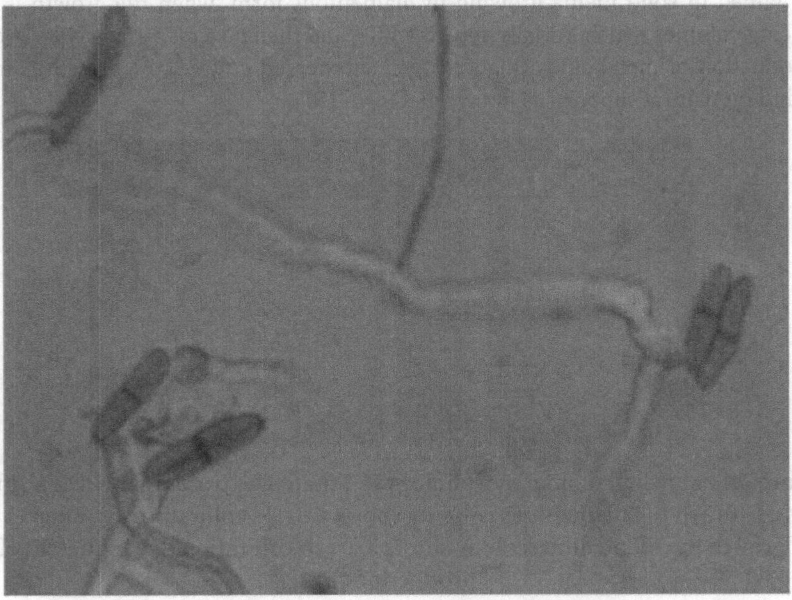

Figure 5. Conidia of *S. maydis* with rounded ends and 1–2 septa.

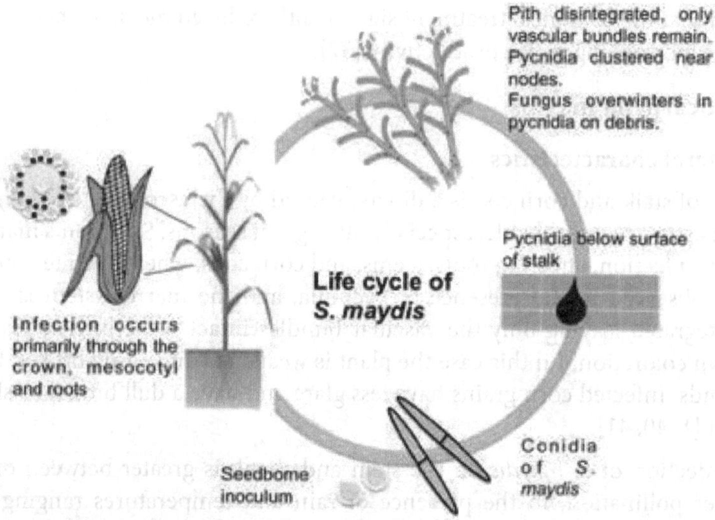

Figure 6. Life cycle of *S. maydis*. Pycnidia that contain the conidia are the principal dissemination source. A sexual stage for this ascomycete has not been described.

The incidence of infected maize by this phytopathogen in the field may range from 1 to 2% or as high as 75 to 80%. This fungus has a worldwide distribution but is mainly found in Guatemala, El Salvador, Belize, Brazil, South Africa, Australia, Asia, and the United States. In the United States this pathogen is the most important causing maize rot [45].

A sexual stage for this ascomycete has not been described. In the laboratory it can be main- tained in solid media growing in filamentous form, when the growth is young, producing colonies which initially appear white, and then take on a green coloration with the production of metabolites (Figure 7). In submerged culture, the growth is in pellet form and on natural supports is in mycelial form [51].

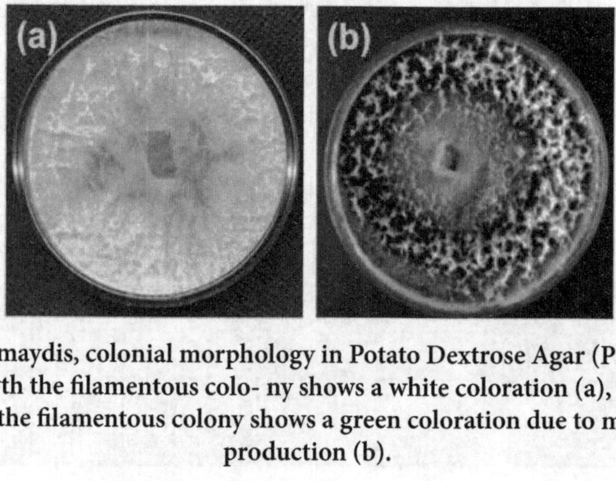

Figure 7. S. maydis, colonial morphology in Potato Dextrose Agar (PDA). After 7 days of growth the filamentous colo- ny shows a white coloration (a), after 15 days of growth the filamentous colony shows a green coloration due to metab- olite production (b).

S. maydis is also an important producer of mycotoxins among which are the diploidiatoxin, chaetoglobosins, and diplonine (Figure 8), all associated with a condition called diplodiosis, a mycotoxicosis characterized by neurological disorders such as ataxia, paralysis, and liver damage in farm animals fed infected corn. The same effect has also been observed in laboratory animals [41, 46–50].

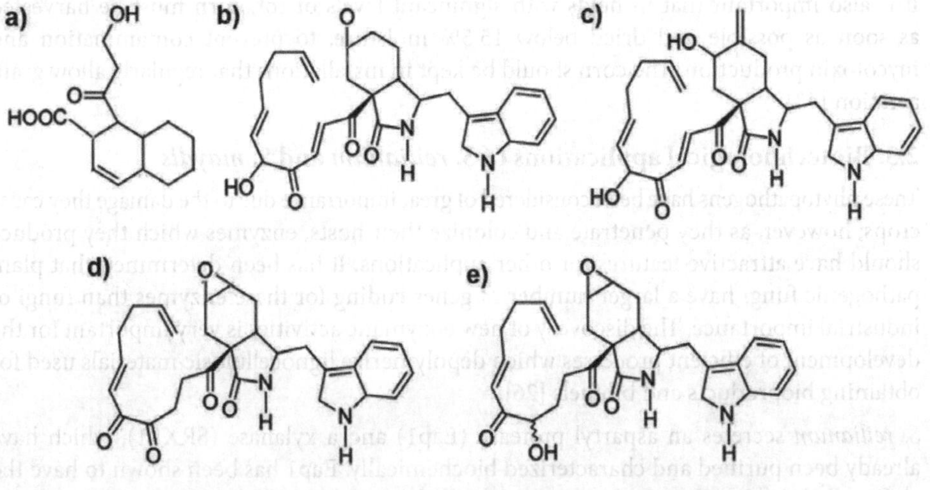

Figure 8. Structures of diplodiatoxin (a), chaetoglobosin K (b), chaetoglobosin L (c), chaetoglobosin M (d), chaetoglobo-sin O (e), diplonine.

2.2.2. Pathogenicity factors

Pathogenicity factors have not been described for this fungus; however, the effect of this ascomycete on the plant must be associated with the production of extracellular enzymes that macerate tissue allowing colonization, as already described for other fungi [26, 44].

S. maydis demonstrated the extracellular hydrolytic activities of acid protease, xylanases, and cellulases when it was grown on solid and liquid fermentation using a synthetic culture medium, as well as when the fungus was grown on the waste of crop maize (i.e., leaf, stem and broken corn), which functioned as inducers for the above mentioned enzymes, suggesting their possible role in the tissue degradation of the host [51].

2.2.3 Control of pathogen

Control of white rot on the stalk and corncob is made by agronomic practices and the use of resistant varieties; either method alter or interrupt the life cycle of the pathogen. There are resistant corn varieties on the market, however, the disease can develop in any hybrid if spore levels are high and climatic conditions are found to favor infection. Proper crop rotation and elimination of infected crop residues can help to reduce the primary inoculum [42].

Another alternative treatment is the application of fungicides, however, the use of these compounds has been reduced due to their high toxicity. In this case, biological control has been an attractive option forming a component of a system of integrated management of disease, consequential t o the decreased use of chemical compounds [38]. The biological control of *S. maydis* has been achieved experimentally, with different strains of actinomycetes demonstrat- ing their potential to become tools for reducing disease

[52, 53]. It has also been reported that strains of bacteria such as *Pseudomonas* spp., *Pseudomonas fluorescens*, *Pantoea agglomerans*, and *B. subtilis* inhibit the development of this fungus for the production of compounds with antifungal activity [54].

It is also important that in fields with significant levels of rot, corn must be harvested as soon as possible and dried below 15.5% moisture, to prevent contamination and mycotoxin production. The corn should be kept in installations that regularly allow grain aeration [43].

2.3. Biotechnological applications of *S. reilianum* and *S. maydis*

These phytopathogens have been considered of great importance due to the damage they cause crops; however, as they penetrate and colonize their hosts, enzymes which they produce should have attractive features for other applications. It has been determined that plant pathogenic fungi have a larger number of genes coding for these enzymes than fungi of industrial importance. The discovery of new enzymatic activities is very important for the development of efficient processes which depolymerize lignocellulosic materials used for obtaining bioproducts and biofuels [26].

S. reilianum secretes an aspartyl protease (Eap1) and a xylanase (SRXL1), which have already been purified and characterized biochemically. Eap1 has been shown to have the ability to degrade proteins in a corn plant and coagulate milk, suggesting it may have potential in the dairy industry, specifically in the production of cheese, or may be used to obtain protein hydrolysates of plant origin. Meanwhile, the xylanase SRXL1 presents interesting biochemical properties, having good stability over a wide range of temperature and pH. This suggests they could be used in the clarification of juices, increasing the performance and enhancing the maceration process, thus reducing the degree of viscosity. They may also improve the digestibility of straw destined as feed for ruminants [28, 29].

S. maydis is capable of producing hydrolytic enzymes such as cellulases, xylanases, and acid protease, into solid and liquid fermentation with different synthetic culture media, where it produces up to two isoforms of either xylanases or cellulases. When it was cultivated using biodegradable supports, it showed three isoforms of xylanases. The most interesting finding is that the fungus produced xylanolytic enzyme extracts free from cellulase activity [51]. These might be used in the paper industry facilitating the release of lignin from paper pulp, thereby reducing the use of chlorine as a bleaching agent, and avoid the degradation of cellulose.

References

1. Bensch M, Staden J. Ultrastructural histopathology of infection and colonization of maize by Stenocarpella maydis (=Diplodia maydis). J Phytopathol. 1992; 136: 312-318. DOI: 10.1111/ j.1439-0434.1992.tb01313.x

2. Martinez C, Roux C, Dargent R. The biological cycle of Sporisorium reilianum f. sp. zeae: An overview using microscopy. *Mycologia*. 2002; 94: 505-514. DOI: 10.2307/3761784

3. Schirawski J, Mannhaupt G, Münch K, Brefort T, Schipper K, Doehlmann G, Di Stasio M, Rössel N, Mendoza-Mendoza A, Pester D, Müller O, Winterberg B, Meyer E, Ghereeb H, Wollenberg T, Münsterkötte M, Wong P, Walter M, Stukenbrock E, Güldener U,

Kahmann R. Pathogenicity determinants in smut fungi revealed by genome comparison. *Science*. 2010; 330: 1546–1548. DOI: 10.1126/science.1195330

4. Ghareeb H, Becker A, Iven T, Feussner I, Schirawski J. *Sporisorium reilianum* infection changes inflorescence and branching architectures of maize. *Plant Physiology*. 2011; 156: 2037-2052. DOI: 10.1104/pp.111.179499
5. Frederiksen R. Head smuts of corn and sorghum. In: Proceedings, 32nd Annual corn and Sorghum Research Conference; Washington DC 1977; pp. 89–105
6. Prom L, Perumal R, Erpelding J, Montes N, Odvody G, Greenwald C, Jin Z, Frederiksen R, Magill C. Virulence and molecular genotyping studies of *Sporisorium reilianum* isolates in sorghum. *Plant dis.* 2011; 95: 523–529. DOI: 10.1094/PDIS-10-10-0720
7. Zhang S, Gardiner J, Xiao Y. Floral transition in maize infected with *Sporisorium reilianum* disrupts compatibility with this biotrophic fungal pathogen. *Planta*. 2013; 237: 1251-1266. DOI: 10.1007/s00425-013-1841-0
8. Zuther K, Kahnt J, Utermark J, Imkampe J, Uhse S, Schirawski J. Host specificity of *Sporisorium reilianum* is tightly linked to generation of the phytoalexin luteolinidin by *Sorghum bicolor*. *Mol Plant Microbe In*. 2012; 25: 1230-1237. DOI: 10.1094/ MPMI-12-11-0314
9. Poloni A, Schirawski J. Host specificity in *Sporisorium reilianum* is determined by dis- tinct mechanisms in maize and sorghum. *Mol Plant Pathol.* 2015; 1-14. DOI: 10.1111/ mpp.12326
10. Matyac C, Kommedahl T. Factors affecting the development of head smut caused by *Sphacelotheca reiliana* on corn. *Phytopathology*. 1985; 75:577-581. DOI: 10.1094/ Phyto-75-577
11. Martinez C, Roux C, Jauneau A, Dargent R. Effect of water potential on the development of an haploid strain of *Sporisorium reilianum* F. sp. *zeae*. *Plant soil*. 2003; 251: 65-71. DOI: 10.1023/A:1022942627449
12. Stoll M, Piepenbring M, Begerow D, Oberwinkler F. Molecular phylogeny of *Ustilago* and *Sporisorium* species (Basidiomycota, Ustilaginales) based on Internal Transcribed Spacer (ITS) Sequences. *Can J Bot*. 2003; 81: 976-984. DOI: 10.1139/B03-094
13. Schirawski J, Heinze B, Wagenknecht M, Kahmann R. Mating type loci of *Sporisorium reilianum*: Novel pattern with three and multiple b specificities. *Eukaryotic Cell*. 2005; 4: 1317-1327. DOI: 10.1128/EC.4.8.1317-1327.2005
14. Williams-alanís H, Pecina-Quintero V, Montes-García N, Arcos-Cavazos G, Zavala-García F, Gámez-Vázquez J. Incidencia de carbón de la panoja *Sporisorium reilianum* (Kühn) Langdon y Fullerton En Híbridos de Sorgo [Sorghum Bicolor (L.) Moench] para grano. Rev Mex Fitopatol. 2009; 27: 36-44.
15. Stienstra W, Kommedahl E, Stromberg E, Matyac C, Windels C, Morgan F. Suppression of corn head smut with seed and soil treatments. *Plant Dis*. 1985; 69: 301-302. DOI: 10.1094/PD-69-301
16. Pradhanang P, Ghimire S. Fungicide management of maize head smut (*Sphacelotheca reiliana*) by seed treatment. *Trop Agric*. 1996; 73: 325-328.

17. Njuguna J. Potential for control of head smut caused by *Sphacelotheca reiliana* in CIMMYT maize germplasm. Sixth East. South. Africa Reg. Maize Conf. NARC Muguga, Nairobi, Kenya. 1999; pp. 67-68.

18. Wright P, Fullerton R, Koolaard J. Fungicide control of head smut (*Sporisorium reilia- num*) of sweetcorn (*Zea mays*). *New Zeal J Crop Hort Sci.* 2006; 34: 23-26. DOI: 10.1080/01140671.2006.9514383

19. Halisky P. Head smut of sorghum, sudan grass, and corn, caused by *Sphacelotheca reiliana* (Ku hn) Clint. *Hilgardia.* 1963; 34:287-304. DOI:10.3733/hilg.v34n08p287

20. Martínez R, Ledesma M. Control químico del carbón de la espiga *Sphacelotheca reili- ana* (Kühn) Clint., del maíz, en el valle de Zapopan, Jalisco. *Rev Mex Fitopatol.* 1990; 8: 68-70.

21. Aquino-Martínez J, Sánchez-Flores A, González-Huerta A, Sánchez-Pale J. Resistance of varieties and hybrids of corn (*Zea mays*) to *Sporisorium reilianum* and grain yield. *Rev Mex Fitopatol.* 2011; 29: 39-49.

22. Little C, Perumal R, Tesso T, Prom L, Odvody G, Magill C. Sorghum pathology and biotechnology. A fungal disease perspective : Part I. Grain Mold, Head Smut, and Ergot. *Eur J Plant Sci Biotechnol.* 2012; 6: 10–30.

23. Quezada A, De León C, Hernández A, Nava C. Evaluation of inoculation methods on maize seeds with *Sporisorium reilianum* f. sp. zeae (Kühn) Langdon & Fullerton. *Rev Mex Fitopatol.* 2014; 31: 80-90.

24. Martinez C, Roux C, Dargent R. Biotrophic development of *Sporisorium reilianum* f. sp. *zeae* in vegetative shoot apex of maize. *Phytopathology.* 1999; 89: 247-253. DOI: 10.1094/PHYTO.1999.89.3.247

25. Zhang S, Xiao Y, Zhao J, Wang F, Zheng Y. Zeae in maize. *Mol Genet Genom.* 2013; 288: 21-37. DOI: 10.1007/s00438-012-0727-3

26. King B, Waxman K, Nenni N, Walker L, Bergstrom G, Gibson D. Arsenal of plant cell wall degrading enzymes reflects host preference among plant pathogenic fungi. *Bio- technol Biofuels.* 2011; 4: 1-14. DOI: 10.1186/1754-6834-4-4

27. Wollenberg T, Schirawski J. Comparative genomics of plant fungal pathogens: The *Ustilago-Sporisorium* Paradigm. *PLoS Pathog.* 2014; 10: 1-3. DOI: 10.1371/journal. ppat. 1004218

28. Mandujano-González V, Arana-Cuenca A, Anducho-Reyes M, Téllez-Jurado A, González-Becerra A, Mercado-Flores Y. Biochemical study of the extracellular aspartyl protease Eap1 from the phytopathogen fungus *Sporisorium reilianum*. *Protein Expr Pu- rif.* 2013; 92: 214-222. DOI: 10.1016/j.pep.2013.10.003

29. Álvarez-Cervantes J, Hernández-Domínguez E, Arana-Cuenca A, Díaz-Godínez G, Mercado-Flores Y. Purification and characterization of xylanase SRXL1 from *Sporisorium reilianum* grown in submerged and solid-state fermentation. *BioResources.* 2013; 84: 5309-5318. DOI: 10.15376/biores.8.4.5309-5318

30. Stromberg E, Kommedahl T, Stienstra W, Windels C, Matyac C. Evaluation of maize hybrids and inbreds for resistance to *Sphacelotheca reiliana*. *Phytopathology.* 1982; 72: 943

31. Pataky JK. Smuts. In: Compendium of maize diseases (White D.G., ed). 3rd. Edition. APS Press. Saint Paul, Minnesota, USA. 33-35 pp.

32. Fullerton RA. A smutty story. *Grower*. 2002; 57: 44-52.

33. Pelmus V, Pelmus A. Aspects of prevention and control of *Sorosporium holci-sorghi* f. *zeae* by chemical seed treatment. 1988; 16: 253-258.

34. Godwin J, Anthony V, Clough M, Godfrey C. ICI A 5504: a novel broad spectrum, systemic ß-methoxyacrylate fungicide. In, Proceedings of the Brighton Crop Protection Conference, Pests and Diseases. 1992, pp. 435-443.

35. Mack J, Baggettt R, Koepsell A, Effects of cultural practices on the incidence of head smut in sweet corn. *HortScience*. 1984; 19: 77–78.

36. Pérez P, Bobadilla M, Velásquez G, Zacatenco A, Espinoza A. Logros y aportaciones de la investigación en maíz de riego en el Valle de Mezquital, Hidalgo. En: *Memorias de veinte años de investigación y desarrollo tecnológico*. Campo Exptl. Pachuca, Hidalgo, INIFAP, Pachuca, México. 2006, p p. 55-65.

37. Mercado-Flores Y, Cárdenas-Álvarez O, Rojas-Olvera V, Pérez-Camarillo P, Leyva- Mir G, Anducho-Reyes A. Application of *Bacillus subtilis* in the biological control of the phytopathogenic fungus *Sporisorium reilianum*. *Biol Control*. 2014; 76: 36-40. DOI: 10.1016/j.biocontrol.2014.04.011

38. Spadaro D, Gullino L. State of the art and future prospects of biological control of postharvest fruit diseases. *Int J. Food Microbiol*. 2004; 91: 185–194. DOI:10.1016/S0168-1605(03)00380-5

39. Compant S, Duffy B, Nowak J, Clément C, Barka A. Use of plant growth promoting bacteria for biocontrol of plant disease: principles, mechanisms of action, and future prospects. *Appl Environ Microbiol*. 2005; 71: 4951-4959. DOI: 10.1128/AEM.71.9.4951-4959.2005

40. McGee C. Maize Diseases: A Reference Source for Seed Technologists. The American Phytopatological Society, St. Paul, MN, USA, 1988.

41. De León C. Enfermedades del maíz. Una guía para su identificación en el campo. Centro Internacional de Mejoramiento de Maíz y Trigo (CIMMYT). 2004, pp. 446-447.

42. Lips P, Mills D. Corn ear rots: *Diplodia* and *Gibberella*. CORN. News Letter. Ohio State Univ. Extension, 2004, 32: 3 p.

43. Vincelli, P. (1997). Ear Rot of Corn Caused by Stenocarpella maydis (= Diplodia maydis). Cooperative Extension Service, College of Agriculture, University of Kentucky, USA.

44. Xia Z, Wu H, Achar PN. Infection and Ultrastructure of Conidia and Pycnidia of *Stenocarpella maydis* in Maize. *J Food Prot*. 2011; 4: 676-680. DOI: 10.4315/0362-028X.JFP-10-347

45. Koehler B. Corn ear rot in Illinois. University of Illinois. Agricultural Experiment Sta- tion. Bull. 639. 1959. 5-6 p.

46. Blaney J, Mannion E, Tuder D, McKenzie A. Examination of Diplodia maydis infected maize for toxicity to chicken and cattle. *Aust Vet J*. 1981; 57: 196–197. DOI: 10.1111/j.1751-0813.1981.tb00513.x

47. Kellerman S, Rabie J, Van Der Westhuizen A, Kriek J, Prozesky L. Induction of diplodiosis: A neuromyco-toxicosis in domestic ruminants with cultures of indigenous and exotic isolates of Diplodia maydis. *Ondersteport J Vet Res.* 1985; 52: 35–42.
48. Rabie J, Kellerman S, Kriek P, Van der Weshuizen C, De Wet J. Toxicity of Diplodia maydis in farm and laboratory animals. *Food Chem Toxicol.* 1985; 23: 349–353. DOI: 10.1016/0278-6915(85)90070-5
49. Rahman F, Rao K, Achar N. Effect of Diplodiatoxin (*Sternocarpella maydis*) on some enzymatic profiles in male and female rats. *Ecotoxicol Environ Saf.* 2002; 52; 267–272. DOI:10.1006/eesa.2002.2200
50. Rogers D, Cannistra C, Gloer B, Wicklow T. Diplodiatoxin, chaetoglobosins, and diplonine associated with a field outbreak of *Stenocarpella ear* rot in Illinois. *Mycotoxin Res.* 2014; 30: 61–70. DOI: 10.1007/s12550-014-0188-0
51. Hernández-Domínguez M, Rios-Latorre A, Álvarez-Cervantes J, Loera-Corral O, Ro- mán-Gutiérrez D, Díaz-Godínez G, Mercado-Flores Y. Xylanases, cellulases, and acid protease produced by *Stenocarpella maydis* grown in solid-state and submerged fer- mentation. *BioResources.* 2014; 9: 2341-2348. DOI: 10.15376/biores.9.2.2341-2358
52. Bressan W. Biological control of maize seed pathogenic fungi by use of actinomycetes. *BioControl.* 2003; 48: 233–240. DOI: 10.1023/A:1022673226324
53. Bressan W, Figueiredo F. Biological control of *Stenocarpella maydis* in maize seed with antagonistic streptomyces sp. isolates. *J. phytopathol.* 2005; 153: 623-626. DOI: 10.1111/j.1439-0434.2005.01014.x
54. Petatán-Sagahón I, Anducho-Reyes A, Silva-Rojas V, Arana-Cuenca A, Téllez-Jurado A, Cárdenas-Álvarez O, Mercado-Flores Y. Isolation of bacteria with antifungal activ- ity against the phytopathogenic fungi Stenocarpella maydis and Stenocarpella mac- rospora. Int. J. Mol. Sci. 2011; 12: 5522-5537. DOI: 10.3390/ijms12095522

Chapter-4

Fungal Growth Control by Chitosan and Derivatives

1. Introduction

The discovery of natural antimicrobial compounds, due to growing consumer demand for food without chemical preservatives, has been focused on numerous studies. In this context, the antimicrobial activity of chitin, chitosan and its derivatives against different types of microorganisms, such as bacteria, fungi and yeasts, has received considerable attention. In this chapter important developments concerning the application of chitosan and its derivatives as antimicrobial compound against fungi and yeasts, assumptions involved in their antimicrobial activity and effects on the quality and storage of fresh vegetables treated with these compounds are described.

The polymers of chitin, chitosan and chito-oligomers have been extensively studied, due to their high potential for applications in food, pharmaceutical, cosmetic and agriculture areas. The applications of these compounds in several areas, especially application of chitosan, is justified by the low cost of production, which is produced from the disposal of processing crustaceans, which are an abundant and renewable source. In general, commercial chitosans are available in the range of molar masses between 50 and 2000 kDa and degree of acetyla- tion (DA) between 0.1 and 0.4 [1].

The polymers chitosan and chitin (Figure 1) are aminoglucopyranans composed of *N*-acetyl- D-glucosamine (GlcNAc) and D-glucosamine (GlcN) residues (Figure 2). The polymers may be distinguished by their solubility in 1% aqueous acetic acid (v/v). Chitin containing ≥40% GlcNAc (FA≥0.4) is insoluble, whereas soluble polymers are named chitosan [2]. Chitosan is composed of three reactive functional groups: an amino group and two hydroxyl groups in the primary and secondary carbons of the positions C-2, C-3 and C-6, respectively [3]. Chemical and biotechnological processes are currently being investigated for the production of chitosan. Industrially chitosan is produced from the alkaline deacetylation of chitin (alka- line hydrolysis), but chitosan can also be obtained from enzymatic deacetylation of chitin, a process investigated in academic studies (Figure 1).

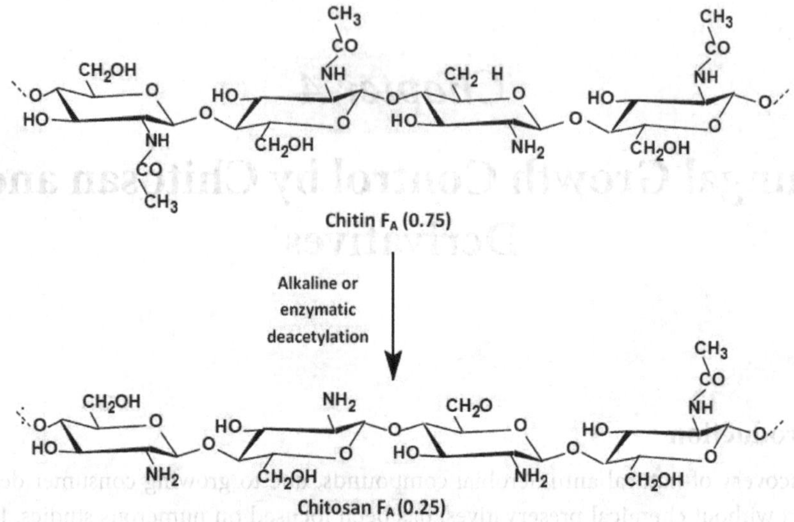

Figure 1. Chemical structures of chitin and chitosan.

Chitin is also widely distributed in fungi, occurring in *Basidiomycetes*, *Ascomycetes* and *Phycomycetes*, where it is a component of cell walls and structural membranes of mycelia, stalks and spores [4]. The amounts vary between traces and up to 45% of the organic fraction, and the remainder is composed mainly of proteins, glucans and mannans [5]. However, not all fungi contain chitin, which may be absent in some species close to other containing chitin as a cell wall component [4].After the discovery of antimicrobial activity of chitosan and its derivatives by Allan & Hadwiger [6], Kendra & Hadwiger [7] and Uchida et al. [8], many researchers have done studies in this field. In this context the antimicrobial activity of chitin, chitosan and its derivatives against different groups of microorganisms such as bacteria, yeast and fungi has received considerable attention.

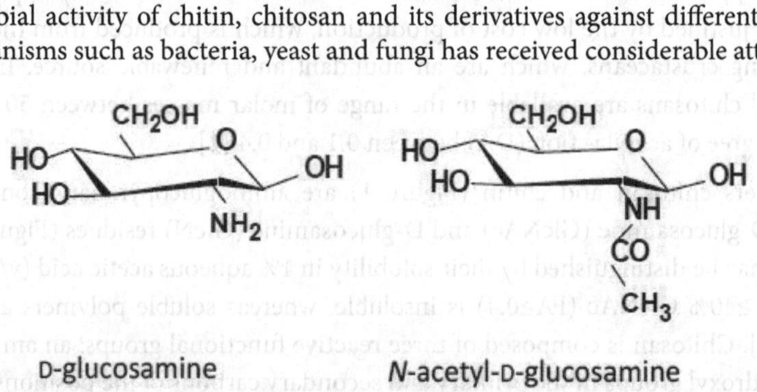

Figure 2. Chemical structures of *N*-acetyl-d-glucosamine (GlcNAc) and d-glucosamine (GlcN) residues.

The antifungal activities of chitosan and its derivatives *in vitro*, preharvest and postharvest studies are reviewed in this chapter. Besides the review of the antifungal activities of chitosan and its derivatives, their abilities to elicit resistance reactions in plants are also reviewed. In addition, chitosan action in the production and viability of fungal spores is reported in this chapter. Finally, the chapter is concluded, with the possible mechanisms suggested in the literature for the antifungal activity of chitosan.

2. In vitro and in vivo antifungal activity of chitosan and its oligomers

The postharvest deterioration due to the action of fungi limits the economic value of stored vegetables. Although fungicides are used extensively in control of postharvest diseases, there is a public interest in reducing these residues in food and in pathogens resistant to fungicides.

Unconventional methods of postharvest pathogens control have been reported in the litera- ture. In addition to studies involving the control of pathogenic fungi by fungicides, other methods have been employed, such as biological control [9], biological control association of $CaCl_2$ [10–12] biological control of association with modified atmosphere [13], postharvest heat treatment [14, 15], heat treatment association and ethanol [16] and chitosan [17, 18].

There is strong evidence that the fungal mycelium growth can be delayed or completely inhibited when chitosan is added to the yeast culture medium. When increasing the chitosan concentration of 0.75 to 6.0 mg x ml^{-1}, El Ghaouth et al. [19] observed a decrease in the radial growth of *Alternaria alternata, Botrytis cinerea, Rhizopus stolonifer* and *Colletrotichum gloeospor- ioides*. The same effect was observed against *Sclerotinia sclerotiorum* by increasing the chitosan concentration of 1 to 4% (w/v) [20]. Other studies showed a linear decrease in the growth of *Rhizoctonia solani* with gradual increase in the concentration of 0.5 to 6.0 mg x ml^{-1} of chitosan [21]. The mycelial growth of *Fusarium solani* f. sp. phaseoli and *F. solani* f. sp pisi was inhibited at minimum concentrations of 12 and 18 mg x ml^{-1} [22, 23]. Complete inhibitions of the fungi *F. oxysporum, R. stolonifer, Penicillium digitatum* and *C. gloeosporioides* were obtained at a concentration of 3% (w/v) [24, 25].

Table 1 lists some studies that evaluated the effects of chitosan in vitro growth of pathogenic fungi.

Table 1. In vitro chitosan effect on fungal growth.

Fungus	Chitosan concentration % (m/v)	Effect	Author
A. alternata, B. cinerea e *R.* stolonifer	0.075 to 0.6	Reduction of radial growth	[26]
Rhizoctonia solani	0.05 to 0.6	Growth reduction	[21]
F. solani f. sp. *phaseoli*	1.2	Complete inhibition	[22]
F. solani f. sp *pisi*	1.8	Complete inhibition	[23]
F. oxysporum, R. stolonifer, Penicillium digitatum e C. gloeosporioides	3	Complete inhibition	[25, 26]
Mucor racemosus	0.2	73% of growth reduction	[27]
A. alternata	0.01 to 0.08	Complete inhibition	[28]
B. cinerea	0.04 to 0.1	Complete inhibition	
P. expansum	0.08	Growth reduction	
R. stolonifer	0.01 to 0.02	Complete inhibition	

Allan & Hadwiger [6] reported that chitosan has a strong antifungal activity against numerous pathogens with the exception of the class *Zygomycetes*, which contains chitosan as the major cell wall component. However, the results of Roller & Covill (1999) [27] demonstrated that the fungus *Mucor racemosus*, whose cell walls are composed of chitosan, was inhibited at a concentration of chitosan 1 g × L^{-1}, contrary to Allan & Hadwiger [6] proposition, which included strains of *Mucor spp.*

No et al. [29] examined the antibacterial activity of chitosans with different molar masses on the growth of gram-positive and gram-negative bacteria. They observed that the growth of gram-positive bacteria was nearly or completely inhibited by all samples of chitosan with different molar mass. On the other hand, for the gram-negative bacteria, the antibacterial activity appeared to increase with decrease of molar mass.

El Ghaouth et al. [30] investigated the effect of chitosan coating (1.0 to 1.5% m/v) in controlling decay of strawberries at 13° C compared to the fungicidal effect of ipridione (Rovral®) and concluded that the coating with chitosan was more effective than treatment with Rovral® fungicide in controlling postharvest decay. The antifungal effects of chitosan on *in vitro* growth of strawberries postharvest have also been studied by El Ghaouth et al. [26]. According to this study, chitosan F_A 0.83 markedly reduced the mycelial growth of the fungi *Botrytis cinerea* and

Rhizopus stolonifer with a great effect at high concentrations. These authors also confirmed the importance of the large number of positively charged groups along the polymer chain due to the fact was observed that *N, O*-carboxymethyl showed lower antifungal activity than chitosan [26]. In an *in vivo* study, El Ghaouth et al. [26] reported that signs of infection were observed in strawberry fruits after five days of storage at 13°C, while the control fruit had shown signs of infection with only one day of storage. After 14 days of storage, chitosan coatings, whose concentration was 15 mg x mL^{-1}, reduced the deterioration strawberries at 60%, caused by the same fungi, and it was observed that the coated fruit usually matured without showing obvious signs of phytotoxicity.

Oliveira Jr. et al. [28] studied the inhibitory effects of fifteen chitosans with different degrees of polymerization (DP) and different DA on the growth rates (GR) of four phytopathogenic fungi (*Alternaria alternata, Botrytis cinerea, Penicillium expansum,* and *Rhizopus stolonifer*) by using a 96-well microtiter plate and a microplate reader. The minimum inhibitory concentra- tions (MICs) of the chitosans ranged from 100 µg × mL^{-1} to 1,000 µg × mL^{-1} depending on the fungus tested and the DP and FA of the chitosan.

Table 2 lists the MICs of chitosan samples that are more effective against fungi *Alternaria alternata, Botrytis cinerea, Penicillium expansum,* and *Rhizopus stolonifer*. Complete inhibition of the fungi *A. alternata, B. cinerea* and *R. stolonifer* and growth reduction of *P. expansum* were obtained with chitosan DP 45 to 1,460 and FA 0.08 to 0.22. Chitosans with smaller values of F_A 0.1 and larger values of DP 3,780 promoted the maximum fungistatic activity against the fungi *B. cinerea* and *A. alternata* [28].

Table 2. MICs of the chitosans with different DP and different FA (fraction of acetylation) against *A. alternata, B. cinerea* and *R. stolonifer* [28].

Fungus	Chitosan samples		MIC ($\mu g \times mL^{-1}$)
	DP	F_A	
A. alternata	190	0.01	200
	320	0.15	400
	121	0.49	800
	3,726	0.10	100
	3,726	0.30	100
	3,850	0.50	300
B. cinerea	190	0.01	800
	3,726	0.10	400
	3,726	0.30	800
	3,850	0.50	800
R. stolonifer	1,383	0.22	200
	45	0.22	200
	1,171	0.08	100
	1,089	0.16	100

Oliveira-Jr et al. [31] have observed that chito-oligosaccharides of DP ≤8 are not notably inhibitory to any of the fungi *A. alternata*, *B. cinerea*, *P. expansum* or *R. stolonifer* and high-DP chito-oligosaccharides (DP ≤12) showed initially inhibitory effects. However, the complete inhibition for all fungi was not obtained by using chito-oligosaccharides. In contrast, as reported by Oliveira-Jr et al. [28], *A. alternata*, *B. cinerea* and *R. stolonifer* were completely inhibited and growth reduction for *P. expansum* was observed by high-DP chitosans (DP 45 to 2,608).

Table 3. Effect of chitosan coating formed on the surface of fruits and vegetables postharvesta immersed in acidic solutions of chitosan or fruits and plants pre harvestb sprayed with chitosan. The fruits, vegetables and plants were inoculated with the respective pathogenic fungi.

Fruit, vegetable or plant	Fungus	Chitosan concentration % (m/v)	Effect (infection reduction %)	Author
Strawberry[a]	*B. cinerea*	1.0 to 1.5	77	[30]
Strawberry[a]	*B. cinerea* and *R. stolonifer*	1.5	60	[19]
Carrots[a]	*Sclerotinia sclerotiorum*	2.0 to 4.0	68	[20]
Cucumber plant[b]	*B. cinerea*	0.1	65	[32]
Strawberry[b]	*B. cinerea*	0.2 to 0.6	45 to 62	[33]
Papaya[b]	*C. gloeosporioides*	1.5	60	[24]

Cuero et al. [34] reported that *N*-Carboxymethyl reduced by 90% of aflatoxin production by *Aspergillus flavus* and *Aspergillus parasiticus*, while the fungal growth was reduced to less than 50%. Chitosan coating was effective in the inhibition of *Sclerotinia sclerotiorum* on carrots (*Daucus carota* L.) whose incidence was significantly reduced from 88 to 28% in the roots, covered with 2 and 4% (m/v) chitosan [20]. Microscopic studies revealed that the mycelium of *Sclerotinia sclerotiorum* exposed to chitosan appeared to be dead and deformed, since the control mycelium was normal.

The control of gray mold caused by *B. cinerea* in cucumber plants by chitosan oligomers was studied by Ben-Shalom et al. [32], who suggested that the primary effect of the oligomers in the control of disease was due to a fungistatic effect on the germination of the conidia of *Botrytis* because chitosan is a positively charged polymer that can prevent binding of conidia some- where. Bhaskara Reddy et al. [33] evaluated the preharvest spraying effect of chitosan on postharvest quality of strawberry and incidence of the pathogen *B. cinerea* and observed that preventive spraying of chitosan was effective in controlling infection of *B. cinerea* on straw- berries.

The relationship between molar mass of chitosans and chito-oligomers and antifungal activity has been analyzed in several studies. Kendra and Hadwiger [23] observed that monomers and dimers of chitosan showed no antifungal activity against *Fusarium solani*, while heptamers had antifungal activity equivalent to chitosan. Uchida et al. [8] reported that a mixture of chito- oligomers with DP 2 to 8 (average of 5 DP) and concentration of 1% (m/v) were inactive against three species of the genus *Fusarium*.

Zhang et al. [35] reported that chito-oligomers with an average DP of 20 inhibited the growths of 16 plant pathogens. Torr et al. [36] suggested that higher antifungal activity against certain fungi may be obtained with chito-oligomers (DP 5, DP 9 and DP 14) when compared to those obtained with chitosan (310 kDa to >375 kDa; DP 1,925 to 2,329). Chitosan acetate and mixtures of chito-oligomers, cited above, were tested against *Leptographium procerum*, *Sphaeropsis sapinea* and *Trichoderma harzianum*. The average GR of *T. harzianum* decreased with the increase of concentration of chitosan acetate and chito-oligomers 0.1 to 0.4% (m/v), which caused an initial period of fungistase and eventually overcome by the fungus. *Sphaeropsis sapinea* and *Leptographium procerum* were more likely to chitosan and chito-oligomer activities than *T. harzianum*, whose growth was inhibited at the concentration of 0.4% (m/v) within 35 days. The antifungal activities of the three mixtures of chito-oligomers were higher at pH 4.0 than at pH 6.0, in which chito-oligomers DP 9 and 14 were more effective against *S. sapinea* and *L. procerum* than the mixture DP 5.

2.1. Action of chitosan in the production of fungal spores

The chitosan effect on spore production by the fungi *F. oxyporum*, *R. stolonifer*, *C. Gloeosporioides*, *A. alternata* f. sp. *lycopersici* and *A. Niger* was analyzed by Bhaskara Reddy et al. [33], Bautista- Baños et al. [24, 25] and Plascencia-Jatomea et al. [37]. The sporulating fungi treated with chitosans are generally lower than the untreated mold. Moreover, in some studies sporulation was completely inhibited when treated with chitosan. However, in some cases it was observed that chitosan stimulates sporulation. The formation of spores of *A. alternata* in the presence of chitosan at concentrations of 100 and 500 $\mu g \times ml^{-1}$ (sub-lethal dose) was significantly higher than the control without chitosan [25, 33].

These authors indicated that high sporulation may have been due to a stress response induced by this polymer.

2.2. Action of chitosan in the fungal spore viability

The viability of fungal spores has been analyzed after treatment with chitosan. Concentrations of 0.75 mg × mL^{-1} reduced the viability of fungal spore germination and tube growth of *B. cinerea* and *R. stolonifer* [19]. In another study, low chitosan concentrations (20-30 μg × mL^{-1}) caused 50% inhibition of germination and 50 μg × mL^{-1} promoted almost complete inhibition of spore germination [32]. These authors also observed that chitosan has promoted the size reduction of the germination tubes, which had a mean size of ~15 μm in the presence of water and 2 μm in the presence of 10 μg × mL^{-1} of chitosan, and both treatments were incubated for 24 hours. Sathiyabama and Balasubramanian [38] evaluated the effect of chitosan concentra- tion on the viability of fungal spores of *Puccinia arachidis* incubated for 4 hours and observed that with increasing chitosan concentration from 100 to 1,000 μg × mL^{-1}, a reduction percentage of the number of germinated spores was observed from 24 ± 3 to 6 ± 0%, respectively. On the other hand, untreated spores with chitosan had a germination of 96 ± 1%.

2.3. Changes in hyphal morphology due to chitosan treatment in some fungal species

Microscopic observations of fungi treated with chitosan showed that the polymer can affect the hyphal morphology. Changes in hyphal morphology, such as excessive mycelial branch- ing, abnormal shapes, swelling and hyphae size reduction, were observed in *Fusarium oxysporum* f. sp. radicis-lycopersici, *R. stolonifer* and *S. sclerotiorum* treated with chitosan [20, 26, 39]. Similarly, chitosan caused morphological changes as large vesicles or empty cells devoid of cytoplasm in the mycelium of *B. cinerea* (Figure 3) and *F. oxyporum* f. sp. *albedinis* [40, 41]. In further studies, the morphology of fungi *C. gloeosporioides*, *R. stolonifer*, *P. digitatum* and *F. oxysporum* treated with chitosan was evaluated. Bautista-Baños et al. [24, 25] reported that area, size and shape of conidia of each fungi tested were affected according to the fungal species and incubation time exposed in the chitosan solutions. Plascencia-Jatomea et al. [37] reported that the morphology of the spores of *A. niger* was also affected when treated with chitosan.

Changes in hyphal morphology due to chitosan treatment in *A. alternata*, *B. cinerea*, *P. expan- sum* and *R. stolonifer* were analyzed by scanning electron microscopy [42]. The micrographs revealed mycelial aggregation and morphological structural change as excessive branching, cell wall swelling and reduction in hyphal length (Figures 4 and 5).

Aggregation, excessive mycelial branching and hyphae size reduction of all fungi treated with chitosan were observed by Oliveira Junior et al. [42]. *A. alternata*, *B. cinerea* and *R. stolonifer* treated with chitosan, besides to have the morphological changes mentioned before, also showed abnormal shapes and swelling in their mycelia (Figures 4 and 5).

The micrographs of *P. expansum* previously treated with chitosan viewed in high magnification of 10,000× showed the chitosan coating formed on surface of the mycelia (Figure 6 A).

The results demonstrated that chitosan acetate was effective in restricting the fungal growth of filamentous fungi [42] by causing a fungistatic inhibition effect as observed by the scanning electron microscopy. In case of *A. alternata*, it was common to observe some spores with germ tube inhibition as shown in Figure 6 B.

Chitosan coating observed on the surface of the mycelia suggested that the fungal growth inhibition could be explained by a direct interaction of chitosan on the fungal cell wall as a consequence of polycationic nature of chitosan. Oliveira-Jr et al. [28] have observed that chitosan samples with low FA (high concentration of free amino groups protonated) and large DP were most effective against the phytopathogenic fungi tested, while chitosan with high FA did not have the ability to inhibit the fungal growth *in vitro*. In another study, OliveiraJr. et al.

[31] have demonstrated that chito-oligosaccharides of eight DP were not notably inhibitory to any of the fungi. On the other hand, higher chito-oligosaccharides (DP 10 and DP 12) showed initially inhibitory effects, which seemed to be more pronounced at a lower F_A.

3. Chitosan as inducer of response mechanisms in plants

Stimulants are substances (oligosaccharides, glycoproteins, peptides and lipids) that can induce defense responses when applied on plant tissue or plant cell culture. Oligosaccharides most studied as inductors are oligomers of glucan, chitin, chitosan and galacturonic acids. When a plant is attacked by a pathogen, fast defense mechanisms are activated in the infected site and various biochemical defense responses occur around the dead cells. Among the biochemical defense responses include the production of reactive oxygen, structural changes in the cell wall, protein accumulation related to defense and biosynthesis of phytoalexins [43].

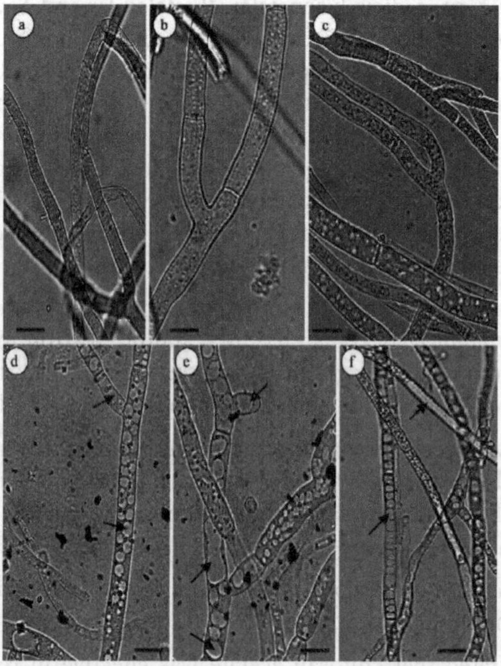

Figure 3. Microscopic structural changes of hyphal fragments of *B. cinerea* in response to the presence of (a) chitosan and (b) control mycelium; (c–f) Mycelia of fungal

cultures grown on PDA containing 1.75% (v/v) of chitosan (Chito- gel®). Bars: 40 μm. Small and large vesicles appeared in the samples treated with chitosan and in some cases the cyto- plasm was free of any organelle (arrows). (Reproduced from Ait Barka et al. [40]. Copyright of Plant Cell Reproduction 2004.)

The stimulatory abilities of chitosan in the natural plant defense responses have been extensively studied. Physiological and biochemical changes that occur in plants due to stimulation by chitosan have been described in several studies [45–53]. Primary physiological changes were observed in plants treated with chitosan, whose openings of the stomata were decreased impeding the fungal access inside the leaf tissues. Lee et al. [44] observed that guard cells of plant leaves produce H_2O_2, which is a mediator compound promoted by chitosan stimulus, which induces a decrease in stomatal openings (Figure 7).

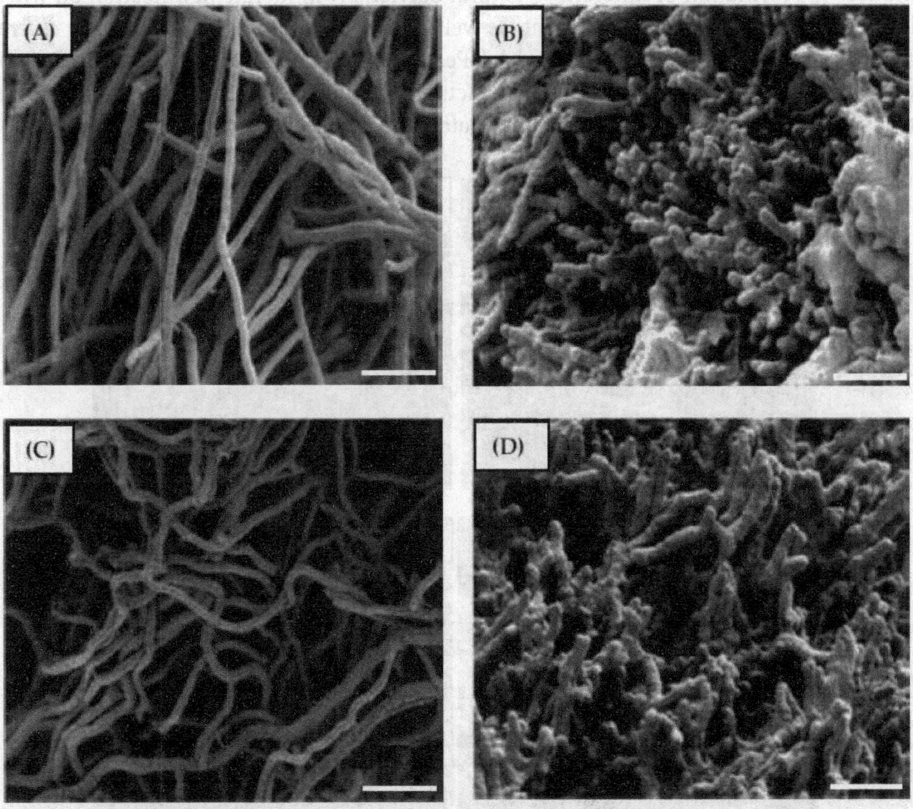

Figure 4. Scanning electron micrographs of mycelia after 5 days of cultivation at 25°C. *Alternaria alternata* (A) control media and (B) medium amended with chitosan (500 μg × mL^{-1}). *Botrytis cinerea* (C) control media and (D) medium amended with chitosan (500 μg × mL^{-1}). Bars = 20 μm. Reproduced from Oliveira Junior et al. [42]. Copyright of Brazil- ian Archives of Biology and Technology 2012.

Chitosan oligosaccharides lignin stimulated accumulation of callose, phytoalexins, and/or protease inhibitors in various plant tissues. The mechanism of action by which induces this lignification chitosan have been studied in different types of plants [46, 54].

Induction of several enzymes related to plant defense process has been studied [45, 46]. These enzymes participate in the initial defense mechanisms and prevent infection by pathogens. Oligomers of chitin and chitosan have been associated with stimulation of other systems involved in resistance as the activity of lipooxygenase and phenylalanine ammonia lyase and the formation of lignin in wheat leaf [45, 46].

The formation of structural barriers on the affected areas by fungi is the most common process response to pathogen invasion. Cell suberization and lignification and other defense processes are stimulated during the process of infection in some organs of plants. Reports describe that chitosan restricted in some cases, the fungal penetration and it induces the formation of different structural barriers. A moderate lignification on wheat leaf, as a result of chitosan treatment, as well as the inoculum of cell walls of *B. cinerea* after 48 and 72 hours were reported by Pearce and Ride [47]. Transmission electron microscopy showed the formation of particular structures and new materials. The main reactions observed in the host cells of tomato roots and leaves treated with chitosan and

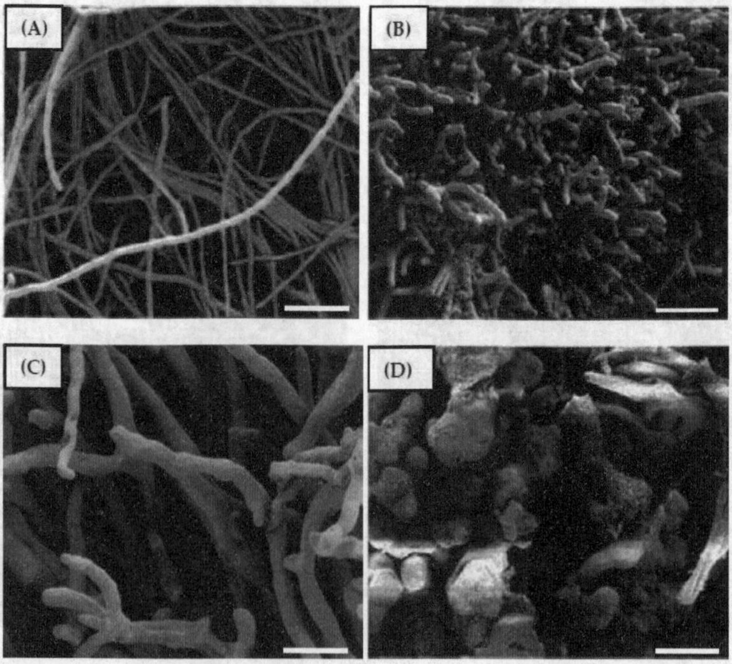

Figure 5. Scanning electron micrographs of mycelia after 5 days of cultivation for *Penicillium expansum* and after 3 days for *Rhizopus stolonifer* at 25 °C. *Penicillium expansum* (A) control media and (B) medium amended with chitosan (500 μg × mL^{-1}). *Rhizopus stolonifer* (C) control media and (D) medium amended with chitosan (500 μg × mL^{-1}). Bars = 20 μm. Reproduced from Oliveira Junior et al. [42]. Copyright of Brazilian Archives of Biology and Technology 2012.

infected by *F. oxyporum* f. sp. *radicis-lycopersici* were as follows: (1) blockage of xylem vessels by an opaque fibrous or granular material or blister- shaped structure; (2) coating the secondary membrane making it thicker and characterized by lesions (3) forming papillae (affixing wall) within the cortex and endothermic tissues [48, 49]. Other reactions of the specific host plant roots of tomato plants treated with chitosan showed deformed epidermal cells [50]. In pepper fruits of bell pepper, structural defense responses were observed only in the first layer of fabric next to broken cells, as thickening of the cell walls, forming spherical and hemispherical protrusions along the cell walls, and blocking of cellular spaces was also observed due to the formation of fibrillar material [51, 52]. Other studies have shown that the combination of two control methods (chitosan and biological control with *Bacillus pumilus*) increased defense reactions of the host plant [53]. Cucumber plants grown in nutrient solutions containing chitosan, and inoculated with *P. aphaniderma- tum*, had similar reactions to those observed in tomato roots treated with chitosan as obstruction of cellular spaces with opaque and fibrillar materials and lastly it was observed the formation of buds along the host cell wall [51].

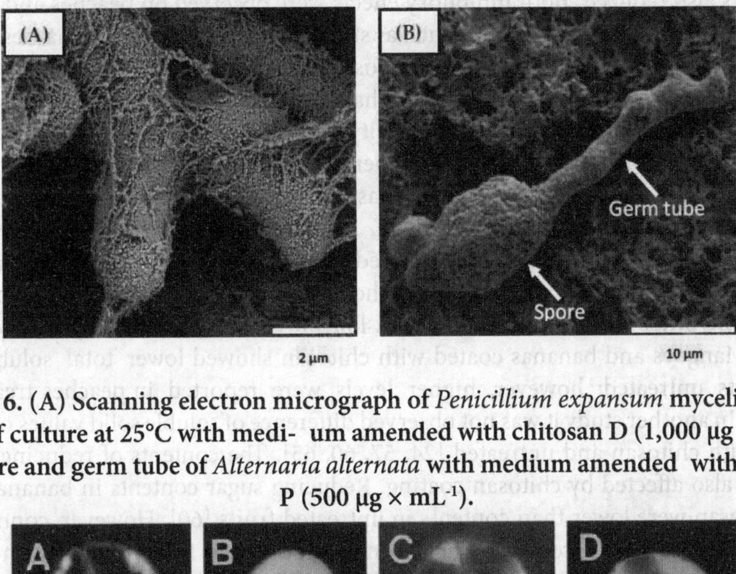

Figure 6. (A) Scanning electron micrograph of *Penicillium expansum* mycelia after 5 days of culture at 25°C with medi- um amended with chitosan D (1,000 µg × mL^{-1}). (B) Spore and germ tube of *Alternaria alternata* with medium amended with chitosan P (500 µg × mL^{-1}).

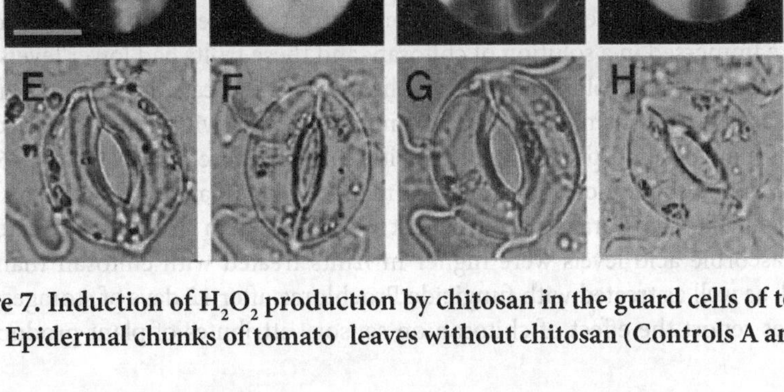

Figure 7. Induction of H_2O_2 production by chitosan in the guard cells of tomato leaves. Epidermal chunks of tomato leaves without chitosan (Controls A and E) or

treatments of 30 minutes with only chitosan (B and F), chitosan and cata- lase (C and G) or with chitosan and ascorbic acid (D and H). Fluorescence microscopy are shown in A–D and optical microscopy are shown in E–H. The bar in A is 10µm and applies to all figures. Reproduced from Lee et al. [44]. Copy- right Plant Physiology (1999).

4. Effect of chitosan on postharvest quality of plant products

Plant products have their shelf life extended when coated with chitosan. Chitosan forms a semi-permeable film that regulates gas exchange and reduces losses by transpiration; therefore, the ripening of the fruit is delayed. Different fruits coated with chitosan, usually have their respiration rates and reduced water losses, among them tomatoes, strawberries, longan, apples, mangoes, bananas and bell peppers [55–60]. The efficacy of chitosan in reducing internal CO_2 production is described in tomatoes and pears [56, 57]. Chitosan coatings associated with storage temperature may be associated with a reduction in CO_2 production. Cucumbers and peppers had lower respiration rates at 13°C than at 20°C [55]. Besides the inhibition of CO_2 resulting of chitosan coating, the ethylene production of fruits is also reduced. Both inhibitory effects were observed on peaches and tomatoes coated with chitosan [56, 61]. Fruits such as strawberries, raspberries, tomatoes, peaches, papaya and other fruits had their firmness loss delayed during storage when treated with chitosan [24, 30, 61]. Sprays of chitosan preharvest at concentrations 2, 4 and 6 g x L^{-1} on strawberry plants did not cause phytotoxicity and the fruits treated with chitosan were firmer than the control fruits [33]. In general, the anthocyanin degradation on fruits treated with chitosan is delayed, which has been demonstrated in lychee, strawberry and raspberry [62–64]. On the other hand, it was observed by El Ghaouth et al. [30] anthocyanin synthesis in strawberries treated with chitosan. Strawberries, tomatoes and peaches treated with chitosan after storage showed higher acidity compared to the control fruits, while other fruits like mangoes and longan had reduced acidity slowly [56, 59, 61, 65]. Mangoes and bananas coated with chitosan showed lower total soluble solids than fruits untreated; however, higher levels were reported in peaches treated with chitosan. In another study it was not observed difference of soluble solid values of papayas treated with chitosan and untreated [24, 57, 60, 65]. The contents of reducing sugar of fruits are also affected by chitosan coating. Reducing sugar contents in bananas treated with chitosan were lower than contents in untreated fruits [60]. However, contradictory reports regarding to the reducing sugar contents of mango fruits treated with chitosan have been described in the literature. A possible explanation for this could be related to the chitosan application method on the surface of the fruit. In the first study, mango fruits were packed in cardboard boxes and covered with chitosan film; in this case the levels of reducing sugars were higher than those of control fruits, while in the second study, mango fruits were immersed in a solution of chitosan, and these fruits had lower levels of reducing sugars than the control fruits [60, 65]. These results indicate that the immersed fruits had decreased metabolism compared to untreated fruits with chitosan. Ascorbic acid content in mangoes and peaches treated with chitosan were also evaluated [61, 65]. In these studies, the content of this vitamin in mango fruits treated with chitosan gradually decreased during the storage period and it was lower than in fruits untreated. But in peaches, ascorbic acid levels were higher in fruits treated with chitosan than in fruits untreated, as well as treated with fungicide Prochloraz after 12 days of storage. Although few studies report the effect of chitosan on sensory attributes of plant products treated with chitosan, some reports showed that flavor and taste remain unchanged. Mangoes and strawberries treated with chitosan had higher scores in the sensory attributes compared to untreated fruit stored for 21 and 15 days, respectively [60, 62]. In other studies, strawberries coated with chitosan and stored for 12 days at 7° C had a slightly bitter taste only on day zero [66].

5. Mode of action of chitosan

Numerous possible mechanisms for the antimicrobial action of chitosan have been proposed, mostly based on the positive charge conferred by protonation of free amino groups at acidic pH, although the exact mechanism of action is still unknown. A polycationic chitosan or oligomer can potentially interact with negatively charged fungal cell membrane components (i.e., proteins, phospholipids), thus interfering with the normal growth and metabolism of the fungal cells [17, 18, 67]. Roller and Covill [27] reported that amino groups in chitosan have the ability to interact with a multitude of anionic groups on the yeast cell wall surface, thereby forming an impervious layer around the cell. Because of its property to form films, chitosan may thus act as a barrier (i.e. anionic groups) and consequently, reducing their availability to a level that will not sustain growth of the pathogen (4). This important property of the polymer chitosan, the ability to protonate at acidic solutions is due to the presence of amines in the molecule that bind to protons as shown in equation (1).

$$Chitosan - NH_2 + H_3O^+ \rightleftharpoons Chitosan - NH_3^+ + H_2O \tag{1}$$

The pKa value of chitosan is approximately 6.3. The chitosan is solubilized when more than 50% of the amino groups are protonated [68]; thus, the solubility of chitosan sharply decreases when the pH increases above 6.0 to 6.5 [18].

Sudarshan et al. [70] and Papineau et al. [71] observed the bacterial agglutination using low concentrations of chitosan lower that 0.2 mg x mL^{-1} probably due to binding of the polycationic polymer to the negatively charged bacterial surface; However, at high concentrations agglu- tination was observed, which according to the authors may be linked to the high number of positive charges that can be formed a positive net charge on the bacterial surface keeping them in suspension.

The interaction between chitosan and the cell can also alter the permeability of the cell membrane. For example, fermentation of yeast used in baking is inhibited by certain cations that act at the cell surface and prevent glucose entry [43]. The interaction between chitosan and *Pythium oaroecandrum* cells was studied by Leuba & Stossel [72], who used a UV technique, and they found that there was considerable release of protein material from the cells at pH 5.8.

Chitosan also acts as a chelating agent that selectively binds to trace metals and thus inhibits toxin production and microbial growth [34].

Liu et al. [73] reviewed the antibacterial activity of chitosan acetate solution against *Escherichia coli* and *Staphylococcus aureus*. The integrity of the cell membrane of both species was investi- gated by determining the release of intracellular materials which absorb at 260 nm. It has been observed a gradual increase in intercellular material suspensions of bacteria treated with chitosan acetate (0.5 and 0.25% w / v) for two hours of monitoring. The authors also noted that according to the results of infrared spectroscopy and thermogravimetric and differential thermogravimetry profiles, there was ionic bond formation between the NH_3^+ group of the chitosan acetate and the phosphoryl group of phosphatidylcholine. This result appears to confirm the electrostatic interaction between chitosan and ethyl phosphatidylcholine, which is a component of bacterial cell membrane.

6. Final Remarks

Chitin, chitosan, derivatives and their oligomers have been widely studied and the existence of great number of scientific papers that have been published in the literature reflects the great potential applications of these polymers, derivatives and oligomers. Considering the global trend of consumer preference for foods without chemical preservatives, the chitosan and other natural compounds have shown to be alternative compounds to control fungi and bacteria, although chemical preservatives are also used extensively in the control of these microorgan- isms, especially the fungicides used in the control of postharvest diseases of fruits. Pre- and postharvest studies of plants, vegetables and fruits have shown that the polymer chitosan has triple effect in the treatment of these; it controls pathogenic microorganisms, it activates various defense responses, inducing and/or inhibiting different biochemical activities during plant-pathogen interaction and it increases the storage time of fresh vegetable due to film formation properties.

References

1. Rege, P. R., Block, L. H. Chitosan processing: influence of process parameters during acidic alkaline hydrolysis and effect of the processing sequence on the resultant chitosan's properties. Carbohydrate Research. 321, n.3-4, 235–245, 1999.
2. Peter, M.G. Chitin and Chitosan from Animal Sources. In: *Biopolymers, Polysaccharides II: Polysaccharides from Eukaryotes*, (A. Steinbüchel series ed.; E. J. Vandamme, S. De Baets and A. Steinbüchel volume eds), 2002a, 6, pp 481–574. Weinheim: Wiley-VCH.
3. Furusaki, E.; Ueno, Y.; Sakairi, N.; Nishi, N.; Tokura, S. Facile preparation and inclusion ability of a chitosan derivative bearing carboxymethyl-β-cyclodextrin. Carbohydrate Polymers. 9, 29–34, 1996.
4. Peter, M.G. Chitin and Chitosan in Fungi. In: *Biopolymers, Polysaccharides II: Polysaccharides from Eukaryotes*, (A. Steinbüchel series ed.; E. J. Vandamme, S. De Baets and A. Steinbüchel volume eds), 2002b, 6, pp 123–157. Weinheim: Wiley-VCH.
5. Roberts, G. A. F. Chitin chemistry. Ed Macmillan, 1992, 350 p.
6. Allan, C.R., Hadwiger, L.A. The fungicidal effect of chitosan on fungi of varying cell composition. Experimental mycology. 3, 285–287, 1979.
7. Kendra, D.F., Hadwiger, L.A. Characterization of the smallest chitosan oligomer that is maximally antifungal to *Fusarium solani* and elicits pisatin formation by *Pisum sati- vum*. Experimental mycology. 8, 276–281, 1984.
8. Uchida, Y., Izume, M., Ohtakara, A. Preparation of chitosan oligomers with purified chitosanase and its application. In: Skjak-Braek, G., Anthonsen, T., Sandford, P. (Eds.). Chitin and chitosan: sources, chemistry, biochemistry, physical properties and applications. Elsevier, London, 373–382, 1989.

9. Melo, I. S. Recursos Genéticos Microbianos. In: Melo, I. S., Valadares-Inglis, M. C., Nass, L. L., Valois, A. C. C. *Recursos Genéticos e Melhoramento-Microrganismos.* Jaguar- iúna: Embrapa Meio Ambiente, 2002. Cap. 1, pp 1–48.

10. Wisniewski M., Droby S., Chalutz E., Eilam Y. Effects of Ca^{2+} and Mg^{2+} on *Botrytis cin- erea* and *Penicillium expansum in vitro* and on the biocontrol activity of *Candida oleophi- la.* Plant Pathology. 44, 1016–1024, 1995.

11. Tian, S. P., Fan Q., Xu, Y., Wang, Y. Effects of *Trichosporon* sp. in combination with calcium and fungicide on biocontrol of postharvest diseases in apple fruits. Acta Bot- anica Sinica. 43, 501–505, 2001.

12. Tian, S. P., Fan, Q., Xu Y., Jiang, A. L. Effects of calcium on biocontrol activity of yeast antagonists against the postharvest fungal pathogen *Rhizopus stolonifer.* Plant Pathology. 51, 352–358, 2002.

13. Qin G., Tian S., Xu Y. Biocontrol of postharvest diseases on sweet cherries by four antagonistic yeasts in different storage conditions. Postharvest Biology and Technol- ogy. 31, 51-58, 2004.

14. Couey, H.M. Heat treatment for control of postharvest diseases and insect pests of fruits. HortScience, Alexandria, 24, n.2, pp198-202, 1989.

15. Fallik, E., Grienberg, S., Gambourg, M., Klein, J. D., Lurie, S. Prestorage heat treat- ment reduces pathogenicity of *Penicillium expansum* in apple fruit. Plant Pathology. 45, 92–97, 1995.

16. Karabulut, O. A.; Gabler, F. M.; Mansour, M.; Smilanick, J. L. Postharvest ethanol and hot water treatments of table grapes to control gray mold. Postharvest Biology and Technology. 34, n.2, p.169–177, 2004.

17. Shahidi, F., Arachchi, J. K. V., Jeon, Y. J. Food applications of chitin and chitosans. Trends in Food Science and Technology. 10, 2, 37–51, 1999.

18. Bautista-Baños S, Hernandez-Lauzardo A.N, Velázquez-Del Valle M.G., Hernández-López M., Bravo-Luna L, Ait Barka E., Bosquez-Molina E., Wilson C.L. Chitosan as a potential natural compound to control pre and postharvest diseases of horticultural commodities. Crop protection. 25, 108–118, 2006.

19. El Ghaouth, E.A., Arul, J., Grenier, J., Asselin, A. Antifungal activity of chitosan on two postharvest pathogens of strawberry fruits. Phytopathology. 82, 398–402, 1992.

20. Cheah, L.H.; Page, B.B.C. New Zealand Journal of Crop and Horticultural Science. 25, 89–92, 1997.

21. Wade, H.E., Lamondia, J.A. Chitosan inhibits *Rhizoctonia fragariae* but not strawberry black root rot. Advanced Strawberry Research. 13, 26–31, 1994.

22. Hadwiger, L.A., Beckman, J.M. Chitosan as a component of pea-*Fusarium solani* interactions. Plant Physiology. 66, 205–211, 1980.

23. Kendra, D.F., Hadwiger, L.A. Characterization of the smallest chitosan oligomer that is maximally antifungal to *Fusarium solani* and elicits pisatin formation by *Pisum sati- vum.* Experimental. Mycology. 8, 276–281, 1984.

24. Bautista-Baños, S., Hernández-López, M., Bosquez-Molina, E., Wilson, C.L. Effects of chitosan and plant extracts on growth of *Colletotrichum gloeosporioides*, anthracnose levels and quality of papaya fruit. Crop Protection. 22, 1087–1092, 2003.

25. Bautista-Baños, S., Hernández-López, M., Bosquez-Molina, E. Growth inhibition of selected fungi by chitosan and plant extracts. Mexican Journal of Phytopathology. 22, 178–186, 2004.

26. El Ghaouth, A., Arul, J.,Asselin, A. Benhamou, N. Antifungal activity of chitosan on post harvest pathogens: induction of morphological and cytological alterations on *Rhizopus stolonifer*. Mycology Research. 96, 769–779, 1992.

27. Roller, S., Covill, N. The Antifungal Properties Of chitosan in laboratory media and apple juice. International Journal of Food Microbiology. 47, 67–77, 1999.

28. Oliveira-Jr, E. N.; El Gueddari, N. E.; Moerschbacher, B. M.; Franco, T. T. Gr Inhibition of phytopathogenic fungi by characterized chitosans. Brazilian Journal of Microbiology. 43, 800–809, 2012.

29. No, H. K., Park, N. Y., Lee, S. H., Meyers, S. P. Antibacterial activity of chitosans and chitosan oligomers with different molecular weights. International Journal of Food Microbiology. 74, 65–72, 2002.

30. El Ghaouth, A., Arul, J., Ponnampalam, R., Boulet, M. Chitosan coating effect on storability and quality of fresh strawberries. Journal of Food Science. 56, 1618–1620, 1991.

31. Oliveira-Jr E.N.; El Gueddari, N.E.; Moerschbacher, B.M.; Peter, M.G.; Franco, T.T. Growth of phytopathogenic fungi in the presence of partially acetylated chitooligosaccharides. Mycopathologia. 166, 163–174, 2008.

32. Ben-Shalom, N., Ardi, R., Pinto, R., Aki, C., Fallik, E. Controlling gray mould caused by *Botrytis cinerea* in cucumber plants by means of chitosan. Crop Protection. 22, 285– 290, 2003.

33. Bhaskara Reddy, M.V., Belcacemi, K., Corcuff, R., Castaigne, F., Arul, J. Effect of pre-harvest chitosan sprays on post-harvest infection by *Botrytis cinerea* and quality of strawberry fruit. Postharvest Biology and Technology. 20, 39–51, 2000.

34. Cuero, R.G., Osuji, G., Washington, A. N-Carboxymethyl chitosan inhibition of afla- toxin production: role of zinc. Biotechnology Letters. 13, 441–444, 1991.

35. Zhang, M., Tan, T., Yuan, H., Rui, C. Insecticidal and fungicidal activities of chitosan and oligo-chitosan. Journal of Bioactive Compatible Polymer. 18, 391–400, 2003.

36. Torr, K.M.; Chittenden, C.; Franich, R.A.; Kreber, B. Advances in understanding bioactivity of chitosan and chitosan oligomers against selected wood-inhabiting fungi. Holzforschung. 59, 559–567, 2005.

37. Plascencia-Jatomea, M., Viniegra, G., Olayo, R., Castillo-Ortega, M.M., Shirai, K. Effect of chitosan and temperature on spore germination on *Aspergillus niger*. Macromolecular Bioscience. 3, 582–586, 2003.

38. Sathiyabama, M., Balasubramanian, R. Chitosan induces resistance components in *Arachis hypogaea* against leaf rust caused by *Puccinia arachidis* Speg. Crop Protection. 17, 307–313, 1998.

39. Benhamou, N. Ultrastructural and cytochemical aspects of chitosan on *Fusarium oxysporum* f. sp. radicis-lycopersici, agent of tomato crown and root rot. Phytopathology. 82, 1185–1193, 1992.

40. Ait Barka, E., Eullaffroy, P., Clément, C., Vernet, G. Chitosan improves development, and protects *Vitis vinifera* L. against *Botrytis cinerea*. Plant Cell Reports. 22, 608–614, 2004.

41. El Hassni, M., El Hadrami, A., Daayf, F., Barka, E.A., El Hadrami, I. Chitosan, antifungal product against *Fusarium oxysporum* f. sp. *albedinis* and elicitor of defence reac- tions in date palm roots. Phytopathologia Mediterranea. 43, 195-204, 2004.

42. Oliveira-Jr, E. N.; Melo, I. S.; Franco, T. T. Changes in hyphal morphology due to chitosan treatment in some fungal species. Brazilian Archives of Biology and Technology. 55, 637–646, 2012.

43. Rabea, E.I., Badawy M.E.T., Stevens, C.V., Smagghe, G., Steurbaut, W. Chitosan as antimicrobial agent: applications and mode of action. Biomacromolecules. 4, 1457–1465, 2003.

44. Lee S; Choi H; Suh S; Doo I S; Oh K Y; Choi E J; Schroeder Taylor A T; Low P S; Lee Y. Oligogalacturonic acid and chitosan reduce stomatal aperture by inducing the evolution of reactive oxygen species from guard cells of tomato and *Commelina communis*. Plant physiology. 121(1), 147–52, 1999.

45. Bohland, C., Balkenhohl, T., Loers, G., Feussner, I., Grambow, H.J. Differential induc- tion of lipooxygenase isoforms in wheat upon treatment with rust fungus elicitor, chitin oligosaccharides, chitosan, and methyl jasmonate. Plant Physiology. 114, 679– 685, 1997.

46. Vander, P., Varum, K.M., Domard, A., El Gueddari, N. E. Moerschbacher, B. M. Comparison of the ability of partially N-acetylated chitosans and chitooligosaccharides to elicit resistance reactions in wheat leaves. Plant Physiology. 118(4), 1353–1359, 1998.

47. Pearce, R.B., Ride, J.P. Chitin and related compounds as elicitors of the lignification response in wounded wheat leaves. Physiology and Molecular Plant Pathology. 20, 119–123, 1982.

48. Benhamou, N., Thériault, G. Treatment with chitosan enhances resistance of tomato plants to the crown and root pathogen *Fusarium oxysporum* f. sp. *radicis-lycopersici*. Physiological and Molecular Plant Pathology. 41, 34–52, 1992.

49. Lafontaine, P.J., Benhamou, N. Chitosan Treatment: an emerging strategy for enhancing resistance of greenhouse tomato plants to infection by *Fusarium oxysporum* f. sp. *radicis-lycopersici*. Biocontrol Science and Technology. 6, 11–124, 1996.

50. Benhamou, N., Lafontaine, P.J., Nicole, M. Induction of systemic resistance to *Fusarium* crown and root rot in tomato plants by seed treatment with chitosan. Phytopathology. 84, 1432–1444, 1994.

51. El Ghaouth, E.A., Arul, J., Wilson, C., Benhamou, N. Ultrastructural and cytochemical aspects of the effect of chitosan on decay of bell pepper fruit. Physiological and Molecular Plant Pathology. 44, 417–432, 1994.

52. El Ghaouth, A., Arul, J., Wilson, C., Benhamou, N. Biochemical and cytochemical as- pects of the interactions of chitosan and *Botrytis cinerea* in bell pepper fruit. Postharv- est Biology and Technology. 12, 183–194, 1997.

53. Benhamou, N., Kloepper, J.W., Tuzun, S. Induction of resistance against Fusarium wilt of tomato by combination of chitosan with an endophytic bacterial strain: ultra-structure and cytochemistry of the host response. Planta. 204, 153–168, 1998.

54. Lesney, M. S. Polycation-like behavior of chitosan on suspension-culture derived protoplasts of slash pine. Phytochemistry., 29(4), 1123–1125, 1990.

55. El Ghaouth, A., Arul, J., Ponnampalam, R. Use of chitosan coating to reduce water loss and maintain quality of cucumber and bell pepper fruits. Journal of Food Proc-essing and Preservation, 15, 359–368, 1991.

56. El Ghaouth, A., Ponnampalam, R., Castaigne, F., Arul, J. Chitosan coating to extend the storage life of tomatoes. HortScience. 27, 1016–1018, 1992.

57. Du, J., Gemma, H., Iwahori, S. Effects of chitosan coating on the storage of peach, Japanese pear, and kiwifruit. Journal of the Japanese Society for Horticultural Sci-ence. 66, 15–22, 1997.

58. Du, J., Gemma, H., Iwahori, S. Effects of chitosan coating on the storability and on the ultrastructural changes of 'Jonagold' apple fruit in storage. Food Preservation Sci- ence. 24, 23–29, 1998.

59. Jiang, Y.M., Li, Y.B. Effects of chitosan coating on postharvest life and quality of longan fruit. Food Chemistry. 73, 139–143, 2001.

60. Kittur, F.S., Kumar, K.R., Tharanathan, R.N., 2001. Functional packaging of chitosan films. Z. Tebensmittelunterssuchung und forshung A. 206, 44–47, 2001.

61. Li, H., Yu, T. Effect of chitosan on incidence of brown rot, quality and physiological attributes of postharvest peach fruit. Journal of the science of food and agriculture. 81, 269–274, 2000.

62. Li, C-F., Chung, Y-C., 1986. The benefits of chitosan to postharvest storage and the quality of fresh strawberries. In: Muzarelli, R., Jeuniaux, C., Graham, G.W. (Eds.), Chitin in Nature and Technology. Plenum Press, New York, pp. 908–913.

63. Zhang, D., Quantick, P.C. Effects of chitosan coating on enzymatic browning and de- cay during postharvest storage of litchi (*Litchi sinensis* Sonn.) fruit. Postharvest Biolo- gy and Technology. 12, 195–202, 1997.

64. Zhang, D., Quantick, P.C. Antifungal effects of chitosan coating on fresh strawberries and raspberries during storage. Journal of Horticultural Science and Biotechnology. 73, 763–767, 1998.

65. Srinivasa, P.C., Baskaran, R., Armes, M.N., Harish Prashanth, K.V., Tharanathan, R.N. Storage studies of mango packed using biodegradable chitosan film. European Food Research and Technology. 215, 504–508, 2002.

66. Devlieghere, F., Vermeulen, A., Debevere, J. Chitosan: antimicrobial activity, interac-tions with food components and applicability as a coating on fruit and vegetables. Food Microbiology. 21, 703–714, 2004.

67. Fang, S.W., Li, C.F., Shih, D.Y.C. Antifungal activity of chitosan and its preservative effect on low sugar candied kumquat. Journal of Food Protection, 56, 136–140, 1994.
68. Rinaudo, M., Pavlov, G., Desbrieres, J. Influence of acetic acid concentration on the solubilization of chitosan. Polymer. 40, 7029–7032. 1999.
69. Varum, K. M., Ottoy, M. H., Smidsrod, O. Water-solubility of partially N-acetylated chitosans as a function of pH: effect of chemical composition and depolymerization. Carbohydrate Polymers. 25, 65–70, 1994.
70. Sudharshan, N. R., Hoover, D. G., Knorr, D. Antibacterial action of chitosan. Food Bi-otechnology. 6, 257–272, 1992.
71. Papineau, A., M. Hoover, D. G., Knorr, D., Farkas, D. F. Antimicrobial effect of wa- ter-soluble chitosans with high hydrostatic pressure. Food Biotechnology. 5, 45–57, 1991.
72. Leuba. S.; Stossel, P. In: Chitin in Nature and Technology; Muzzarelli, R. A. A., Jeu-niaux, C., Eds.; Plenum: New York, 1985; p 217.
73. Liu, H., Du, Y., Wang, X., Sun, L. Chitosan kills bacteria through cell membrane dam-age. International Journal of Food Microbiology. 95, 147–155, 2004.

Chapter-5

Investigating the Effects of Plant Essential Oils on Post- Harvest Fruit Decay

1. Introduction

An essential oil is a concentrated hydrophobic liquid containing volatile aromatic compounds from plants. Essential oils are also known as volatile oils or ethereal oils. An oil is "essential" in the sense that it carries a distinctive essence of the plant. It does not form a distinctive category for either medical, pharmacological, or culinary purposes.

Essential oils have several applications in the manufacture of perfumes, cosmetics, and soaps. In the food industry, essential oils can be used for flavouring, and for adding scents to incense. The use of some essential oils as alternative antimicrobial agents, replacing chemical treat- ments, has attracted considerable interest from post-harvest scientists in recent years. Most essential oils and their single constituents have been reported to inhibit post-harvest fungi both *in vitro* and *in vivo*. The use of these volatile compounds has garnered a great deal of interest recently.

There is no doubt that essential oils have important pharmacological properties, which result in their wide use in pharmaceutical practices. The first investigation into the antimicrobial activity of essential oils was a study by Buchholtz [1] who found that thymol has higher growth inhibitory properties on bacteria having been cultivated in a tabac decoction than phenol, which was previously considered to be the best systematic substance for use as surgical antiseptic [2]. Therefore, a great deal of research has been carried out aimed at studying the antimicrobial activities of several essential oils that have already been registered in the pharmacological industry.

The fungitoxic activity of essential oils may be due to synergism among their components, since most of this activity has been reported to be enhanced when combined [3]. Bioactivity in the vapor phase of essential oils was recognized as a characteristic that makes them attractive for use as possible fumigants for the protection of stored product [3–6].

This chapter will address different aspects relating to plant essential oils and fungal pathoge- nicity ranging from the chemical use in controlling plant pathogenic fungi as well as discussing the main sources of essential oils, and their biological activity. Furthermore, this chapter will provide an insight into the mode of action of single constituents of different essential oils, using different case studies.

2. Chemical control of plant pathogenic fungi

Collectively, fungi and fungal-like organisms cause more plant diseases than any other group of plant pathogens, with over 8,000 species shown to cause disease. The importance of fungi as agents of plant and human diseases, producers of industrial and pharmacological products, and as decomposers, has prompted scientists worldwide to study their biology. Some of the world's great famines and human suffering can be blamed on plant pathogenic fungi [7].

Chemical disease control employs the use of chemicals that are generally toxic to pathogens and characterized by their specific effect, such as many commercial fungicides and antibiotics. Basically, any chemical agent must be effective and act safely, without harming any live organism and must also have minimal or no effect on the environment, microflora and soil.

Chemical fungicides are generally used in the control of fungal diseases.

Recently, growing public concern over the health and environmental hazards associated with the increased levels of chemical pesticides and the lack of approval for the renewal of some of the most effective active molecules, has led to the development of safe, alternative, and natural methods of post-harvest disease control [8]. To minimize the development of pest resistance as a result of the overuse of chemical fungicides, it was advisable to use a variety of fungicides belonging to different chemical groups. Throughout the 1980s and 1990s, public concern about the use of agricultural chemical pesticides grew as a result of the high risk of poisoning to all living organisms, beneficial micro-flora and micro-fauna, and the contamination of food products. The risk from chemical pesticides is due to the possibility that they may be absorbed through the skin, inhaled or ingested through consumption of contaminated plants.

To overcome the risks of chemical pesticides, some precautions should be taken with respect to the handling of the chemicals and the management of agriculture practices. Environmental concerns focus mainly on protectant fungicides such as copper and sulfur sprays, which have the potential to affect a broad range of organisms if they are washed off leaves and then accumulate in the soil or are washed into waterways. A lot of countries have already started educational programs for farmers to reduce the overuse of chemical pesticides, and hence increase the productivity of several economic crops.

Recently, there has been a great interest in the use of essential oils and plant extracts as possible natural substitutes for conventional synthetic pesticides. This may be attributed mainly to ecosystem pollution and pesticide resistance in pests, insects and fungal pathogens.

3. Plant essential oils

Essential oils are volatile compounds produced in many species of plant. These oils are thought to play a role in plant defense mechanisms acting against phytopathogenic microorganisms [9–13]. Subsequently, plant essential oils were subjected to pharmacological studies and, later, to numerous and frequent tests of their antimicrobial activity.

Several methods have been used to evaluate the *in vitro* antimicrobial activity of different essential oils. Among the most common methods are agar diffusion tests, serial broth or agar dilution tests, and vapor phase tests [14].

Further antimicrobial tests comprise the following:

1. **Various kill-time studies:**

This test deals with determining the activity of a compound relative to the activity of phenol, after 15 min (phenol coefficient).

2. **Killing time determination:**

This test deals with determining the exact time needed for the complete inhibition of the target organism after contact with the test compound, using contaminated silk threads.

3. **LD-50 and recording of growth curves:**

This test records growth curves and determines the amount of a compound effectively inhibiting the growth of 50% of test organisms.

4. **Poisoned food techniques:**

Tests in which the delay of microbial growth is determined in the presence of growth inhibitors.

5. **Spore germination:**

This test is suitable for use with fungi, especially during short contact time studies.

6. **Measuring of metabolic CO_2:**

This test monitors the presence or absence of the growth of yeast, or visualizes growth using indicators such as sulfur salts from sulfur supplemented cow milk as a growth medium, 2,3,5- triphenyltetrazolium chloride, or p-iodo-nitrophenyltetrazolium violet.

3.1. Main sources of plant essential oils

1. Family Lamiaceae

Sage

Sage is considered to be the main genus among the *Lamiaceae* family, which consists of about 900 species widely distributed in the temperate, subtropical and tropical regions all over the world but especially in the Mediterranean region, central Asia, central and South America, and in southern Africa. Globally, the best known species of the family, used in both traditional and modern medicine are *Salvia officinalis*, *S. fruticosa* and *S. divinorum*. Another important plant is oregano, considered to be the most valued species worldwide. About 60 plant species were listed within this common name. The majority of oregano species belong to the *Lamiaceae* and *Verbenaceae*.

Oregano

Oregano is a perennial herb; its flowers are purple, 3–4 mm long, and produced in erect spikes. It is sometimes called wild marjoram, and its close relative *Origanum majorana* is known as sweet marjoram. Many subspecies and strains of oregano have been developed by humans over the centuries for their unique flavors or other characteristics. It is an important culinary herb, used for the flavor of its leaves, which can actually be more flavorful when dried than fresh. It has an aromatic, warm and slightly bitter taste, which can vary in intensity.

The antifungal and antibacterial activity of oregano essential oil against a number of plant pathogens has been reported by Adebayo et al. [15] These pathogens include *Aspergillus niger* v. Tieghem, *A. flavus* Link, *A. ochraceus* Wilhelm, *Fusarium oxysporum* Snyder and Hansen, *Penicillium* sp. L., *Pseudomonas aeruginosa* Schroter ATCC 2730, *Staphylococcus aureus* Rosenbach ATCC 6538, *F. solani* var. *coeruleum* (Martius) Saccardo, *Clavibacter michiganensis* S., *Phytophtora infestans* Mont., *Sclerotinia sclerotiorum* Lib. and *Xanthomonas vesicatoria* D.

Thymes

Thyme is an evergreen herb with culinary, medicinal and ornamental uses. The most common variety is *Thymus vulgaris*. Thyme belongs to the genus *Thymus* of the mint family (*Lamiaceae*) and is a relative of the oregano genus *Origanum*.

The essential oil of *T. capitatus* Hoffm. & Link, displayed antifungal activity in stored foods and inhibited the growth of both *B. cinerea* and *Monilinia fructicola* [16]. *T. vulgaris* showed antifungal activity against some post-harvest fungal pathogens such as *B. cinerea*, *P. italicum*, *P. citrophthora* and *Rhizopus stolonifer* [10].

Marjoram

Marjorana hortensis (*Lamiaceae*), commonly known as marjoram, is a perennial herb or under-shrub with sweet pine and citrus flavors. It has a long history of medicinal and culinary use.

Camele et al. [10] reported that the last two essential oils of *Lamiaceae* family showed antifungal activity against *P. citrophthora* and *R. stolonifer*. *M. hortensis* showed antifungal activity against *C. acutatum* and *B. cinerea*, and antibacterial activity against two strains of G+ve (*Bacillus megaterium* and *C. michiganensis*) and five strains of G-ve (*Escherichia coli, X. campestris, B. mojavensis, P. savastanoi* and *P. syringae* pv. *phaseolicola*) (Elshafie et al., data not published).

2. Family Verbanacae

Vervain

Verbena officinalis (*Verbenaceae*) commonly known as vervain, is a perennial medicinal plant which grows natively in Europe. It is widely naturalized outside its native range, for example in North America. It has been used traditionally as folk medicine in some countries, and recently was reported to act as an anticancer agent. Despite its widespread uses, the mecha- nisms of the pharmacological actions of the herb are still unclear.

The antimicrobial activity of vervain essential oil was reported recently by Elshafie et al. [12] who found that this essential oil significantly reduced (*in vivo*) the brown rot lesion diseases of peach caused by *M. laxa*, *M. fructicola* and *M. fructigena*.

3. Magnoliaceae

Magnolia

Magnolia is a large genus of about 210 flowering plant species in the family *Magnoliaceae*. Its trees are very ancient and the flowers and oil have been used in many cultures around the world.

M. liliflora essential oil showed potential *in vitro* and *in vivo* antifungal effects against *B. cinerea, Colletotrichum capisci* (Syd.) E.J. Butler & Bisby, *F. oxysporum* Snyder and Hansen, *F. solani, P. capisci* Leonian, *Rhizoctonia solani* (Cooke) Wint. and *S. sclerotiorum* [17].

The essential oil of *M. liliflora* also showed a potential *in vivo* antifungal effect against *P. capsici*, and this activity could be attributed to its constituents: α-terpineol, α-bourbonene, β- caryophyllene, 2-β-pinene, α-humulene, farnesene, and caryophyllene oxide components [17].

3.2. Antimicrobial activity of plant essential oils

Numerous essential oils have shown an antifungal effect against several post-harvest pathogens, e.g. *B. cinerea* Pers. [18–20], *Aspergillus* spp. [21–23], *Fusarium* spp. [24,25], *Penicillium* spp. [26], *R. stolonifer* (Ehrenb.: Fr.) Vuill. [27,28], *C. gloeosporioides* Penz. [29]. The antifungal effect is attributed mainly to the inhibition of both mycelial growth and spore germination. This hypothesis suggests that the expression of disease will be restricted by the impeding of the initial infection and the subsequent mycelial spread beyond the infection site [30].

Clove oil is an essential oil extracted from the clove plant, *Syzygium aromaticum* (L.) Merr. & Perry, and has been reported to act as a bioactive substance — especially its active component monoterpene eugenol [31] — against *B. cinerea, M. fructigena* Honey, *P. expansum* Link and *Phlyctema vagabunda* Desm. in apples. Carvacrol (one of the major constituents of oregano essential oil) is a phenol that was reported to show a high inhibition of mycelium growth in *Neofabraea alba* (E.J. Gutrie) Verkley on apples [32].

Essential oils from basil (*Ocimum basilicum* L.), fennel (*Foeniculum sativum* Mill.), lavender (*Lavandula officinalis* Chaix), marjoram (*O. majorana* L.), oregano (*O. vulgare* L.), peppermint (*Mentha piperita* L.), rosemary (*Rosmarinus officinalis* L.), sage (*Salvia officinalis* L.), savory (*Satureja montana* L.), thyme (*T. vulgaris* L.) and wild mint (*Mentha arvensis* L.) showed a potentially significant antifungal activity higher than that available from chemical treatments in post-harvest treatments against *B. cinerea* and *P. expansum* on apples [3].

The different efficacies of essential oils are due to the differing fungitoxic properties of each single active constituent, as well as the synergic effect [3,32]. This hypothesis suggests that the possible phytotoxic effects of essential oil treatments may be due to the same active compo- nents existing in each essential oil. The length of storage may also negatively influence the antifungal activity of the essential oil treatments. Therefore, treatments using essential oils should only be used for short storage times, or they should be repeated after a defined time period has elapsed, depending on the fruit cultivar in question [3]. The selection of an essential oil for a post-harvest treatment must be based mainly on the characteristics of the fruit, desirable storage time and decay.

The antifungal activity of essential oils could be enhanced by the method of application. The potential of using essential oils by dipping or spraying to control post-harvest decay has already been examined in fruit and vegetables [33–35]. The combination of various post- harvest treatments may improve the efficacy of controlling post-harvest pathogens [30,36,37].

Essential oils from thyme (*T. capitatus* L.), spearmint (*M. spicata* L.) and anise (*Pimpinella anisum* L.) exhibited inhibitory effects on the development of *M. fructicola* (G.Winter) Honey [36,37]. The essential oil of Lemon myrtle (*Backhousia citriodora* F.Muell.) has been reported to have antifungal activity against *M. fructicola*, and has been shown to have a strong antimicro- bial activity, mainly with respect to its potential application as a topical pharmaceutical product. Its main constituent, citral, had also been reported to exhibit a fungitoxic effect against a range of post-harvest pathogens [38].

Mechanism of antifungal activity of essential oils

Apart from the positive effect of each of the chemical constituents of different essential oils, several different studies have indicated that there seems to be a synergetic effect between the individual chemical constituents. This synergism in the aromatic plants components functions to make them more effective and reduces the developing resistance of any pathogenic fungi.

In particular, some single constituents such as carvacrol, γ-terpinène and p-cymene become more effective when they are combined together and act synergistically [15].

On the other hand, p-cymene is efficient facilitator of the transport of carvacrol across cell wall components and the cytoplasmic membrane of pathogenic fungi. Thymol, eugenol and carvone are widely used in the control of several fungi, particularly those which contaminate various important economic crops [13].

Another hypothesis suggested by Soylu et al. [39,40], is that the observed diameter reduction and lyses of the hypha wall, may be attributed to the enzymatic reactions within the essential oil which act to regulate synthesis of the wall. Furthermore, the lipophilic properties of the above mentioned single components might have the ability to degrade the plasma membrane, and thus to increase the permeability of the cytoplasm.

The following will give an insight into some case studies relating to the antimicrobial activity of some Mediterranean plant essential oils:

1st case study: *Biochemical characterization of oregano essential oil*;

2nd case study: *Antimicrobial activity of oregano, thyme and vervain essential oils*; 3rd case study: *In vivo antifungal activity of thyme and vervain essential oils*;

4th case study: *In vitro antifungal activity of the main components of vervain, thyme and oregano essential oils*;

5th case study: *In vitro and in vivo antifungal activity of single constituents of oregano essential oil.*

3.3. Case studies

First case study

Biochemical characterization of oregano essential oil

Mancini et al. [11] studied the biochemical characterization of *O. vulgare* ssp. *hirtum* essential oil from the Southern Apennines (Italy). *O. vulgare* is composed mainly of phenolic compounds belonging to the carvacrol/thymol chemotype. The possible inhibitory fungicidal activity of *O. vulgare* was determined against *M. laxa* (Aderh. & Ruhland)

Honey, *M. fructigena* and *M. fructicola* as follows: prepare different concentrations of each oil + potato dextrose agar (PDA) + (0.2 %) with 0.2 % Tween 20 and 250, 500, and 1000 ppm of each essential oil under study, then pour 14 ml of PDA+ oil in Petri dish.

Following that, completely dry off the preparation under laminar flow, 0.5 cm disc from the studied fungi 96 h old was inoculated in the centre of each Petri dish. All plates were incubated at 22°C for 96 h in the absence of light, and the diameter of any fungal mycelium growth was measured in mm. PDA plates + Tween 20 without oils were inoculated with the same fungi as a control. Fungitoxicity was expressed as the percentage of growth inhibition (PGI) and calculated according to Zygadlo et al. [41] as follows:

$$PGI\left(\%\right) = 100 \times \left(GC - GT\right) / GC$$

Where GC represents the average diameter of fungi grown in PDA (control); GT represents the average diameter of any fungi cultivated on the treated PDA containing the essential oil.

The oils tested have shown antifungal activity against *M. laxa*, *M. fructigena* and *M. fructicola* (Table 1) [11] and have neither shown any phytotoxic activity against germination and initial radicle elongation of *Sinapis arvensis* L., *Phalaris canariensis* L., *Lepidium sativum* L. and *Raphanus sativus* L., nor have they shown any haemolysing effect against the cell membrane of bovine erythrocytes [11].

Table 1. Antifungal activity of O. vulgare essential oils from three different localities.

		Inhibition Percentage (PGI %)		
Treatments		*M. laxa*	*M. fructigena*	*M. fructicola*
O. vulgare (M)	**1000 ppm**	100.0 ± 0.0a	100.0 ± 0.0a	100.0 ± 0.0a
	500 ppm	79.6 ± 2.21b	69.2 ± 1.67c	54.3 ± 2.65c
	250 ppm	67.9 ± 1.10c	58.0 ± 0.83d	14.3 ± 6.19e
O. vulgare (MP)	**1000 ppm**	100.0 ± 0.0a	100.0 ± 0.0a	100.0 ± 0.0a
	500 ppm	74.2 ± 3.31c	62.1 ± 3.35cd	45.6 ± 2.65c
	250 ppm	48.4 ± 2.21d	38.4 ± 3.35e	20.6 ± 2.65de
O. vulgare(**SGP**)	**1000 ppm**	100.0 ± 0.0a	100.0 ± 0.0a	100.0 ± 0.0a
	500 ppm	92.9 ± 1.10a	86.9 ± 1.68b	85.00 ± 1.77b
	250 ppm	46.8 ± 4.42d	43.1 ± 3.34e	25.6 ± 0.88d

Where: M: Mandia sample; MP: Marconia di Pisticci sample; SGP: San Giovanni a Piro sample. Values followed by the same letter in each vertical column are not significantly different according to Tukey test at $P < 0.05$, data are expressed as mean (SDS), R=3.

Second case study

Antimicrobial activity of oregano, thyme and vervain essential oils

Adebayo et al. [15] reported that oregano essential oil has antimicrobial activity against a number of plant pathogens such as *A. niger* van Tieghem, *A. flavus* Link, *A. ochraceus*

Wilhelm, *F. oxysporum*, *F. solani* var. *coeruleum* (Martius) Saccardo, *Penicillium* spp., *P. aeruginosa* Schroter ATCC 2730, *S. aureus* Rosenbach ATCC 6538, *C. michiganensis* (Smith) Davis, *P. infestans* Mont, *S. sclerotiorum* Lib. and *X. gardneri* Dowson. In addition, different species of oregano such as *O. vulgare* and *O. syriacum* L. showed antifungal activity against *B. cinerea* [33,34].

Thyme and vervain essential oils were obtained from *T. vulgaris, V. officinalis* L. have shown *in vitro* potential fungicidal activity against four causal agents of post-harvest orange fruit rot: *B. cinerea*, *P. italicum*, *P. citrophthora* (R.E. Sm. & E.H. Sm.) Leonian and *R. stolonifer* (Ehrenb.: Fr.) Vuill. [10].

Third case study

In vivo antifungal activity of thyme and vervain essential oils

Elshafie et al. [12] studied the *in vivo* antifungal activity of both thyme oil and vervain oil against post-harvest brown rot disease of peach. They reported that thyme oil was mainly composed of *o*-cymene (56.2%), while the main components of vervain oil were citral (44.5%) and isobornyl formate (45.4%) [10]. In addition, both oils were evaluated for their *in vivo* antifungal activity against the post-harvest pathogen *Monilinia* on peach fruits. All tested peach fruits of cv. "Springcrest", were not treated by either pre- or post-harvest chemical pesticides, and were sterilized superficially with a 2 % sodium hypochlorite solution and were later washed with sterile distilled water. They were finally air dried before being inoculated with the above mentioned three phytopathogenic fungi at room temperature by injuring the fruit surfaces and inoculation with about 10 μL of fungal suspension of 10^6 spore/ mL. Spore suspensions were prepared by adding two loopfuls of each fresh fungal myceli- um (7–10 days old) to 10 mL of sterile distilled water before each of the suspensions was filtered and the concentration adjusted by serial dilution in sterile distilled water. One day after inoculation, each single fruit group was sprayed with different concentrations of thyme essential oil at 250 and 500 ppm, or at 500 and 1000 ppm of vervain essential oil. Each experiment was repeated twice. Four fruits, after being wounded with a sterile needle, were sprayed only with sterile distilled water and used as negative control, while four fruits for each oil concentration were not inoculated and were used instead as a control to determine the possible phytotoxicity. All the fruit series were kept in a moist chamber at a high relative humidity (about 95%) for 4 days at room temperature before being observed for eventual appearance of symptoms. The fungitoxicity effectiveness was expressed as the diameter of the brown rot lesion in mm on fruit with respect to control. Both oils studied showed a promising fungicidal effect *in vivo* on the post-harvest diseases of peach fruits such as brown rot lesion caused by *M. laxa*, *M. fructigena*, and *M. fructicola* (Figure 1,12).

Fourth case study

In vitro antifungal activity of the main components of vervain, thyme and oregano essential oils Camele et al. [42] investigated the biological activity *in vitro* of the main components "β- fellandrene, β-pinene, camphene, carvacrol, citral, *o*-cymene, γ-terpinene and thymol" extracted from three Mediterranean aromatic plants (*V. officinalis*, *T. vulgaris* and *O. vulgare*), against five etiological agents of post-harvest fruit decay, *B. cinerea*, *P. italicum*, *P. expansum*, *P. citrophthora* and *R. stolonifer*. The possible fungistatic or fungicidal activity of each studied essential oil components was determined as follows: (A) putting single 3-mm-thick and 0.5- cm-diameter PDA plugs containing fungal mycelium onto the central part of the surface of Petri dishes containing PDA pre-treated with different concentrations (50, 150 or 250 ppm) of each tested single component dissolved in 0.2% Tween-20, (B) dropping, under axenic conditions, 10 μL aliquots of single suspensions containing 1×10^4 conidia/mL of the single target microorganism species onto the central part of surface of Petri dishes containing PDA prepared with the same percentage of Tween-20 and the single three component concentration. Results showed that citral exhibited a fungicidal action against *P. citrophthora*; carvacrol and thymol showed a fungistatic activity against *P. citrophthora* and *R. stolonifer*. Thymol showed fungicidal activity against *P. italicum*. Citral and carvacrol at 250 ppm, and thymol at 150 ppm all stopped the growth of *B. cinerea*.

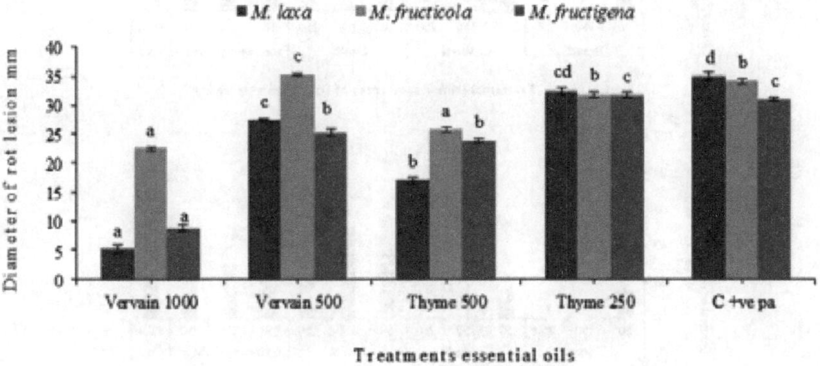

Where: C+ve pa: is positive control where the fruits were challenged only with pathogens.

Figure 1. In vivo antifungal activity of thyme and vervain oil against postharvest brown rot disease of peach caused by M. laxa, M. fructicola and M. fructigena. Bars with different letters indicate means values significantly different at P < 0.05 according to Tukey test. Data are expressed as mean of three replicates ± SD.

Fifth case study

In vitro and in vivo antifungal activity of the single constituents of oregano essential oil

Elshafie et al. [13] evaluated the antifungal effect of the single components of *O. vulgare* ssp. *hirtum* essential oil against the post-harvest pathogens *M. laxa*, *M. fructigena* and *M. fructicola in vitro* and *in vivo*. The chemical characterization of *O. vulgare* reported by Mancini et al. [11] explained that it contains five main single components: carvacrol, thymol, linalool, citral and *trans*-caryophyllene [11].

In vitro antifungal activity

The possible fungicidal activity of the above five standards was determined according to the method of Soylu et al. [40]. Three-mm-thick and 0.5-cm-diameter PDA plugs, axenically taken from the peripheral portion of basic colonies, were inoculated onto the central part of PDA Petri dishes pre-treated with different concentrations of each single component (50, 150 or 250 ppm) dissolved in 0.2% Tween 20. All plates were incubated at 22°C for 96 h in the absence of light. Negative controls comprised either PDA plates without any treatments, or PDA plates treated only with 0.2% Tween 20. The antifungal activity was expressed by measuring the diameter of any mycelium growth in mm [42].

Carvacrol and thymol have exhibited the highest activity during *in vitro* tests against all tested post-harvest *Monilinia* pathogens (Figure 2,13). Citral showed moderate antifungal activity, lower than that of carvacrol and thymol. Linalool and *trans*-caryophyllene showed slight antifungal activity against all studied pathogens. On the other hand, thymol showed fungitoxic inhibition, whereas carvacrol and citral showed fungistatic activity.

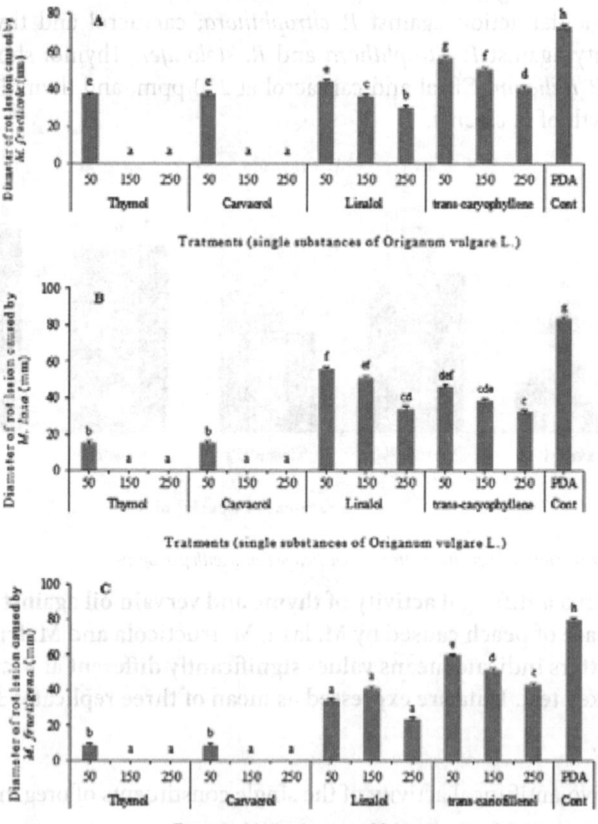

Where: 50, 150 and 250 are the concentrations of each single substance in ppm; PDA is potato dextrose agar.

Figure 2. In vitro antifungal activity of the four single substances of O. vulgare oil against M. laxa, M. fructicola and M. fructigena. Bars with different letters indicate means values significantly different at P < 0.05 according to Tukey test. Data are expressed as mean of three replicates ± SE.

In vivo antifungal activity

The bioactive treatments which exhibited *in vitro* activity were selected for evaluation of their *in vivo* activity against three *Monilinia* species causing brown rot of peach fruits, following the method of Hong et al. [43]. Tested peach fruits cv. "Springcrest", were not treated with either pre- or post-harvest chemical pesticides, and were superficially sterilized with 2 % sodium hypochlorite solution, then later washed with sterile distilled water, before they were finally, air dried and inoculated with the above mentioned three phytopathogenic fungi at room temperature. Each inoculum was performed by injuring the surface of the fruits with a sterile needle and then adding 10 µl of fungal suspension containing 10^6 spore /ml. Liquid fungal cultures were prepared by adding two 3-mm-thick and 0.5-cm-diameter (4 days old) fungal discs to 150 ml of sterilized potato dextrose broth (PDB) medium. They were then incubated at 22 °C for 7–9 days. One day after inoculation, each fruit group was sprayed with an emulsion containing different concentrations of each single component at 150 or 500 ppm, dissolved in 0.2% Tween 20. The negative control composed of three groups of fruit sprayed only with sterile distilled water, whereas the positive control composed of three groups of fruits inoculated only with *Monilinia* isolates. The severity of symptoms induced by infection of the single *Monilinia* isolates was determined by measuring the diameter of brown rot lesions in mm after 3–5 days of incubation at room temperature (16–24 °C).

Carvacrol and thymol have shown a promising inhibition of the brown rot of peach fruits caused by *M. laxa, M. fructicola* and *M. fructigena in vivo* especially at a dose of 500 ppm (Figure 3,13). In particular, carvacrol showed the highest significant antifungal activity against *M. fructicola*.

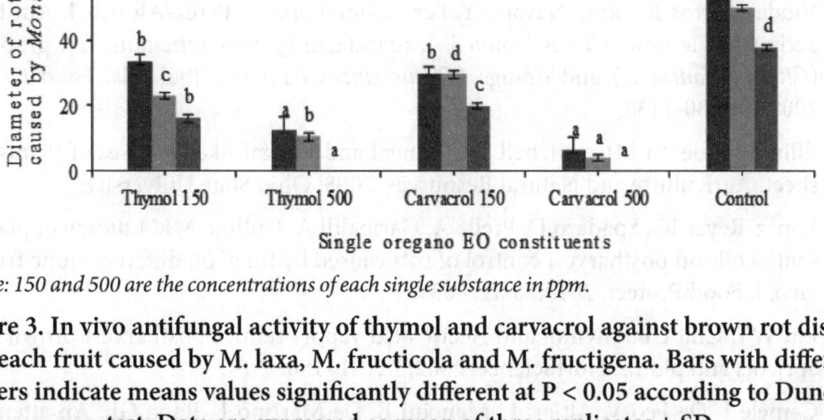

Where: 150 and 500 are the concentrations of each single substance in ppm.

Figure 3. In vivo antifungal activity of thymol and carvacrol against brown rot disease of peach fruit caused by M. laxa, M. fructicola and M. fructigena. Bars with different letters indicate means values significantly different at P < 0.05 according to Duncan test. Data are expressed as mean of three replicates ± SD.

4. Conclusions

Plant essential oils are one several promising environmentally friendly alternatives to conventional synthetic pesticides used to control several fungi and fungus-like organisms such as plant pathogens, food contaminants and decays. Most of the essential oils and their formula- tions mentioned in this review showed a high efficacy — either via direct contact or through incorporation into nutrient media — against different fungal pathogens at very low concen- trations when used either *in vitro* or *in vivo*. Among the essential oils used, oregano, thyme and vervain and their active constituents such as carvacrol, thymol, linalool, citral and their isomers, effectively inhibited mycelial growth and spore germination through fungistatic and/ or fungicidal actions. The active ingredients of the essential oils and their isomers can be used effectively as seed and soil treatments — controlling most of the post-harvest decay fungi — and in plant and food protection.

References

1. Buchholtz L. Antiseptika und Bakterien. Archivexperim. Pathologie Pharmakologie. 1875;4:1–82.
2. Ashhurst AP. The centenary of Lister (1827–1927). A tale of sepsis and anti-sepsis. Ann. Med. History. 1927;9:205–211.
3. Lopez-Reyes JG, Spadaro D, Gullino ML, Garibaldi A. Efficacy of plant essential oils on post-harvest control of rot caused by fungi on four cultivars of apples *in vivo*. Flavour Fragr. 2010;25:171–177.
4. Hidalgo PJ, Ubera JL, Santos JA, LaFont F, Castellanos C, Palomino A, Roman M. Es- sential oils in *Culamintha sylvatica* Bromf. ssp. *ascendens* (Jordan) P.W. Ball wild and cultivated productions and antifungal activity. J. Essent. Oil Res. 2002;14:68-71.
5. Kordali S, Kotan R, Mavi A, Cakir A, Ala A., Yildirim A. Determination of the chemical composition and antioxidant activity of the essential oil of *Artemisia dracunculus* and of the antifungal and antibacterial activities of Turkish *Artemisia absinthium, A. dracunculus*, *Artemisia santonicum*, and *Artemisia spicigera* essential oils. J. Agric. Food Chem. 2005;53:9452-9458.
6. Viuda-Martos M, Ruiz-Navajas Y, Fernandez-Lopez J, Perez-Alvarez J. Antifungal activity of lemon (*Citrus lemon* L.), mandarin (*Citrus reticulata* L.), grapefruit (*Citrus paradise* L.) and orange (*Citrus sinensis* L.) essential oils. Food Control. 2008;19:1130-1138.
7. Ellis SD, Boehm MJ., Mitchell TK. Fungal and Fungal-like Diseases of Plants. Fact sheet, Agriculture and Natural Resources. 2008; Ohio State University.
8. Lopez-Reyes JG, Spadaro D, Prelle A, Garibaldi A, Gullino ML. Efficacy of plant essential oils on postharvest control of rots caused by fungi on different stone fruits in vivo. J. Food Protect. 2013;76:631–639.
9. Liu WT, Chu CL. Thymol and Acetic acid vapors reduce postharvest brown rot of apricots and plums. HortScience. 2002;37:151–156.
10. Camele I, De Feo V, Altieri L, Mancini E, De Martino L, Rana GL. An attempt of post- harvest orange fruit rot control using essential oils from Mediterranean plants. J. Med. Food. 2010;13:1515–1523.
11. Mancini E, Camele I, Elshafie HS, De Martino L, Pellegrino C, Grulova D, Vincenzo De Feo. Chemical composition and biological activity of the essential oil of *Origanum vulgare* ssp. hirtum from different areas in the southern Apennines (Italy). Chem. & Biodiver. 2014;11(4):639-651.
12. Elshafie HS, Mancini E, Camele I, Martino LD, De Feo V. *In vivo* antifungal activity of two essential oils from Mediterranean plants against post-harvest brown rot disease of peach fruit. Indus Crop Prod. 2015a;66:11–15.

13. Elshafie HS, Mancini E, De Martino L, Pellegrino C, De Feo V, Camele I. Antifungal activity of some constituents of *Origanum Vulgare* L. essential oil against post-harvest disease of peach fruit. J. Med. Food. 2015b;18(8):929-934.
14. Handbook of essential oils, Science, technology, and applications. Edited by: K. Hüsnü Can Baser and Gerhard Buchbauer. CRC Press. Taylor & Francis Group.
15. Adebayo O, Dang T, Bélanger A, Khanizadeh S. Antifungal Studies of Selected Essential Oils and a Commercial Formulation against *Botrytis Cinerea*. J. Food Res. 2013;2(1):217-226.
16. Tsao R, Zhou T. Antifungal activity of monoterpenoids against post-harvest pathogens *Botrytis cinerea* and *Monilinia fructicola*. J. Essent. Oils Res. 2000;12:113–121.
17. Bajpai VK, Kang SC. *In Vitro* and *In Vivo* Inhibition of Plant Pathogenic Fungi by Essential Oil and Extracts of *Magnolia liliflora* Desr. J. Agr. Sci. Tech. 2012;14:845-856.
18. Wilson CL, Solar JM, Ghaouth EL, Wisniewski ME. Rapid evaluation of plant extracts and essential oils for antifungal activity against Botrytis cinerea. Plant Dis. 1997;81:204–210.
19. Walter M, Jaspers MV, Frampton CM, Eade K, Stewart A. Control of *Botrytis cinerea* in grape using thyme oil. Australas. Plant Pathol. 2001;30:21–25.
20. Bouchra C, Achouri M, Idrissi LM, Hmamouchi M. Chemical composition and antifungal activity of essential oils of seven Moroccan *Labiatae* against *Botrytris cinerea* Pers. Fr. J. Ethnopharmacol. 2003;89:165–169.
21. Dubey NK, Kishore N, Shing SN. Antifungal properties of the volatile fraction of *Melaleuca leucodendron*. Trop. Agric. 1988;60:227–228.
22. Rahman MU, Gul S. Mycotoxic Effects of Thymus Serpyllum Oil on the Asexual Re-production of *Aspergillus* Species. J. Essent. Oils Res. 2003;15:166–171.
23. Shashi SK, Patra M, Shukla AC, Dikshit A. Use of essential oil as botanical–pesticides against post-harvest spoilage in *Malus pumilo* fruits. Biocontrol. 2003;48:223–232.
24. Vaughn FS, Spencer FG. Antifungal activity of natural compounds against thiabendazole-resistant *Fusarium sambucinum* strains. J. Agric. Food Chem. 1994;42:200–203.
25. Daferera DJ, Ziogas BN, Moschos G. The effectiveness of plant essential oils on the growth of *Botrytis cinerea*, *Fusarium* sp. and *Clavibacter michiganesis* subsp. *michiganen- sis*. Crop Prot. 2003;22:39–44.
26. Arras G, Aggabio M, Piga A, D'Hallewin G. Fungicide effect of volatile compounds of *Thymus capitatus*. Acta Hort. 1995; 379:593–600.
27. Thompson DP. Effect of essential oils on spore germination of *Rhizopus stolonifer* and *Aspergillus* species. Mycologia. 1993;78:482–485.
28. Reddy MVB, Angers P, Gosselin A, Arul J. Characterization and use of essential oil from *Thymus vulgaris* against *Botrytis cinerea* and *Rhizopus stolonifer* in strawberry fruits. Phytochemistry. 1998;47:1515–1520.
29. Palhano FL, Vilches TT, Santos RB, Orlando MT, Ventura AJ, Fernandes PM. Inactivation of *Colletorichum gloeosporioides* spores by high hydrostatic pressure combined with citral or lemongrass essential oils. Int. J. Food Microbiol. 2004;95:61–66.

30. Lazar-Baker EE, Hetherington SD, Ku VV, Newman SM. Evaluation of commercial essential oil samples on the growth of post-harvest pathogen *Monilinia fructicola* (G. Winter) Honey. Lett. in Appl. Microbiol. 2011;52:227–232.

31. Amiri A, Dugas R, Pichot AL, Bompeix G. *In vitro* and *in vivo* activity of eugenol oil (*Eugenia caryophylata*) against four important post-harvest apple pathogens. Int. J. Food Microbiol. 2008;126(1-2):13-9.

32. Neri F, Mari M, Brigati S, Bertolini P. Control of Neofabrea alba by plant volatile compounds and hot water. Post-harvest Biol. Technol. 2009;51:425-430.

33. Smid EJ, Dewitte Y, Gorris LG. Secondary plant metabolites as control agents of post-harvest *Penicillium* rot on tulip bulbs. Post-harvest Biol. Technol. 1995;6:303-312.

34. Dixit SN, Chandra H, Tiwari R, Dixit V. Development of a botanical fungicide against blue mould of mandarins. J. Stored Prod. Res. 1995;31:165-172.

35. Arrebola E, Sivakumar D, Bacigalupo R, Korsten L. Combined application of antagonist *Bacillus amyloliquefaciens* and essential oils for the control of peach post-harvest diseases. Crop Prot. 2009;29(4):369–377

36. Feng W, Zheng XD, Chen JP, Yang Y. Combination of cassia oil with magnesium sul- phate for control of post-harvest storage rots of cherry tomatoes. Crop Protect. 2008;27:112-117.

37. Dutta PK, Tripathi GK, Mehrotra J. Perspectives for chitosan based antimicrobial films in food applications. Food Chem. 2009;114:1173-1182.

38. Tsao R, Zhou T. Antifungal activity of monoterpenoids against post-harvest pathogens *Botrytis cinerea* and *Monilinia fructicola*. J. Essent. Oils Res. 2000;12:113–121.

39. Soylu S, Yigitbas H, Soylu EM, Kurt S. Antifungal effects of essential oils from oregano and fennel on *Sclerotinia sclerotiorum*. J. Appl. Microbiol. 2007;103:1021-1030.

40. Soylu EM, Kurt S, Soylu S. *In vitro* and *in vivo* antifungal activities of the essential oils of various plants against tomato grey mould disease agent *Botrytis cinerea*. International J. Food Microbiol. 2010;143(3):183-189.

41. Zygadlo JA, Guzman CA, Grosso NR. Antifungal properties of the leaf oils of *Tagetes minuta* L. and *T. filifolia* lag. J. Essent. Oil Res. 1994;6:617–621.

42. Camele I, Altieri L, De Martino L, De Feo V, Mancini E, Rana GL. *In vitro* Control of Post-Harvest Fruit Rot Fungi by Some Plant Essential Oil Components. Int. J. Mol. Sci. 2012;13:2290-2300.

43. Hong C, Michailides TJ, Holtz BA. Effects of wounding, inoculum density, and bio- logical control agents on post-harvest brown rot of stone fruits. Plant Dis. 1998;82:1210–1216.

Chapter-6

Synthetic Peptides as an Alternative Tool for the Diagnosis of Cryptococcosis

1. Introduction

Cryptococcosis is an important systemic mycosis that threatens the lives of humans and animals. It manifests primarily through respiratory system diseases and meningoencephalitis. Cryptococcosis is among the emergent fungal infections with significant morbi-lethality, and it is the fourth most frequent cause of opportunistic infection in human immunodeficiency virus (HIV)-positive patients. The disease is caused by two species of the genus *Cryptococcus*: *Cryptococcus neoformans* and *Cryptococcus gattii* [1, 2].

C. neoformans has a worldwide distribution and is responsible for the high morbi-lethality in immunocompromised individuals with AIDS. In contrast, infections with *C. gattii* are preva- lent in tropical and subtropical climate regions, and *C. gattii* primarily attacks immunocom- petent hosts. However, *C. gattii* has also emerged in countries with temperate climates, e.g., Canada (Vancouver) and the U.S. Northwest, which demonstrates that the fungus may adapt to new environments and cause surges of infection in animals and humans [3–7].

Annually, AIDS-related cryptococcal meningitis is responsible for approximately 15% of the mortality in these individuals [8]. Sub-Saharan Africa has the largest rate of coinfection with *Cryptococcus* in these patients, and *Cryptococcus* is the most common cause of meningitis in adults [9, 10]. Recent estimates indicate an incidence of approximately 10,000 cases per year of cryptococcal meningitis in Latin America [11].

2. Etiological agents

C. neoformans and *C. gattii* are basidiomycetes in the asexual phase, appear as round cells, though they are occasionally ovoid, are isolated or budding, and encased by a mucopolysac- charide capsule. *C. neoformans* is the anamorphic phase of *Filobasidiella neoformans*, and *C. gattii* is the anamorphic phase of *Filobasidiella bacillispora* [12, 13].

C. neoformans was originally divided into two strands: var. *neoformans* (serotypes A, D, and a hybrid AD) and var. *gattii* (serotypes B and C). In 2002, *C. neoformans* var. *gattii* was recognized as a distinct species, *C. gattii*. Furthermore, previously observed phenotypic differences, newer molecular studies, and the sequencing of the fungus' genome were helpful to detect significant genetic variations between serotypes A and D and to distinguish serotype A as a new strand, *C. neoformansvar. grubii* [14, 15].

However, this classification becomes difficult, as significant divergences between serotypes are frequently observed at the molecular level [16]. Serotype limits do not entirely coincide with genetic groupings; therefore, serotyping is not regarded as a reliable technique for differentiating strands of *Cryptococcus* [17].

A series of molecular studies were conducted, including polymerase chain reaction (PCR) fingerprinting and amplified fragment length polymorphism (AFLP) analysis of the orotidine monophosphate pyrophosphorylase (*URA5*) gene and analysis of the phospholipase (*PLB1*) gene by restriction fragment length polymorphism (RFLP). As a result of these analy- ses, the yeasts were classified into the following nine molecular types: VNI (AFLP1) and VNII (AFLP1A and AFLP1B) (*C. neoformans* var. *grubii*, serotype A), VNIV (AFLP2) (*C. neoformans* var. *neoformans*, serotype D), VNIII (AFLP3) (Hybrid, serotype AD), VNB (only one isolated in Botswana) and VGI (AFLP4), VGII (AFLP6), VGIII (AFLP5) and VGIV (AFLP7), all correspond- ing to serotypes B and C, indicating that they evolved independently and in parallel [18–20].

C. gattii genotype VGII was responsible for approximately 95% of the cryptococcosis infections that occurred in the Island of Vancouver, Canada, and in the U.S. [3, 4, 21]. Genetic studies with multilocus sequence typing (MLST), which uses the presence of virulence genes to determine subgroups, have shown that the VGIIa and VGIIb subtypes are responsible for the majority of cryptococcosis cases. Another subtype, VGIIc, which is also virulent, has emerged in Oregon, U.S. and is now, together with subtype VGIIa, contributing to the rise of the disease in that region [4, 22].

C. neoformans genotype VNI and *C. gattii* genotype VGI are regarded as the primary agents of cryptococcosis worldwide. However, in Latin America, the distribution and occurrence of *C. gattii* types differ from those in other continents [17, 19, 23, 24]. In Brazil, the *C. gattii* genotype VGII type is responsible for infections in immunocompetent hosts in the Northern (N) and Northeastern (NE) regions.

Genotype VGI is endemic in Australia and has also been described in Papua New Guinea, Asia, and southern California. The VGIII and VGIV genotypes are found less frequently, with the VGIII type isolated in the Ibero-American regions and in India and type VGIV recorded in South Africa and in the U.S. [19, 25–29].

3. Natural history of the disease

Cryptococcosis is a systemic mycosis with a pulmonary gateway, which is caused by infection with either *C. neoformans* or *C. gattii.* Infection is initiated after inhalation of the fungus' infective propagules, which are the basidiospores, or desiccated yeasts, that are dispersed in the environment [27, 30]. In the majority of cases, inhalation produces a self-limited asymp- tomatic pulmonary infection, which is dependent on the host's immune response, the inocu- lum's size, and the microorganism's virulence. Residual focuses with viable fungal elements can be established, and these can be reactivated after a number of years. At times, it may mimic tuberculosis, with nodular lesions with no calcification and eventual cavitation. Other presentations include a controlled mass similar to neoplasia, and at times, the disease manifests as pneumonia that can evolve into acute respiratory failure.

The pulmonary form is the second most frequent form and attacks 35.7% of HIV-negative patients [31, 32].

Once in the lung, *C. neoformans* or *C. gattii* transit through the blood-brain barrier (BBB) to reach the central nervous system (CNS), causing meningoencephalitis and, in the most serious forms of the disease, brain cryptococcomas. The fungus shows high tropism in the CNS, which is attributed to the optimal concentration of existing nutrients in the cerebrospinal fluid that can be assimilated by the fungus (thiamine, glutamic acid, glutamine, dopamine, carbohydrates, and minerals), as well as to the lack of complement system activity in the cerebrospinal fluid (CSF) and the poor or absent inflammatory response activity of the brain tissue [33–35]. Del Poeta *et al.* [1] recently found that despite the various mechanisms proposed to be responsible for this neurotropism, it is still unknown as to how exactly this occurs during the infection in humans. The *C. gattii* incubation period associated with outbreaks is known to be short. It is believed that this strand can be more aggressive prior to its dissemination into the CNS. In the disseminated form, one can observe the development of cutaneous infections in the form of papulae, pustules, or subcutaneous nodules. In addition, there are cases of primary cutaneous infection with no dissemination, as well as infections in other organs [34, 36, 37].

4. Virulence factors

The pathogenic species of the genus *Cryptococcus* has a number of well-elucidated determi- nants of virulence. Some notable determinants include the ability to produce melanin, extracellular enzymes, the capacity to survive and proliferate at 37°C (thermal tolerance), and to escape oxidative damage caused by the host and the presence of a large polysaccharide capsule [30, 38, 39].

The polysaccharide capsule is composed of 90 to 95% glucuronoxylomannan (GXM), 5% galactoxylomannan (GalXM), and approximately 1% mannoproteins (MPs) [40, 41]. It is regarded as one of the most important virulence factors for *C. neoformans* and *C. gattii*, and its cell components accumulate in body fluids, thus serving as targets for diagnosis. The mecha- nism proposed to explain the contribution of the capsule to virulence is its capacity to inhibit phagocytosis, inactivate components of the complement system, induce apoptosis, and regulate cytokine synthesis. Inside the macrophages, *Cryptococcus* spp. releases and accumu- lates fragments of the polysaccharide, which is shown to be cytotoxic to the macrophage, causing dysfunction or cell death [42–44].

5. Diagnosis of cryptococcosis

The laboratory diagnosis of cryptococcosis is based upon a number of principles: the demon- stration of the yeast in the clinical material, the isolation of the yeast in the culture followed by biochemical tests for the final identification, anatomic-pathological examination, and research into circulating antigens. Several biological materials may be used for the identifica- tion of fungal infection, e.g., serum, plasma, blood, tissue, and CSF, which is the major biological material used for the diagnosis of cryptococcal infection in the CNS [45].

The direct research of the fungus can be accomplished using CSF, sputum, bronchial washing, cutaneous-mucosal lesion pus, urine, macerates of biopsy tissue, prostatic secretion, blood, and bone marrow biopsy specimens. Clinical samples analyzed with

India ink indicate the presence of the capsulated yeasts (Figure 1). This method is fast and low-cost but is not very sensitive and cannot distinguish between species. Due to the high yeasts load found in samples from AIDS patients, the sensitivity of this method may reach 80% for cryptococcal meningitis, whereas in immunocompetent individuals, this sensitivity may be as low as 30–72% [46–48]. In addition, the success of this technique is dependent upon the expertise of the microbiologist, and there are reports in the literature of false negatives in 20–30% of the results from infections with *C. neoformans* or *C. gattii* due to a deficient capsule or the low fungal charge of the agent in the CSF. This is primarily an issue with the initial cases, when the diagnosis is fundamental. Examination of *C. neoformans* or *C. gattii* in tissues is carried out with specific dyes, e.g., mucicarmine and silver [49].

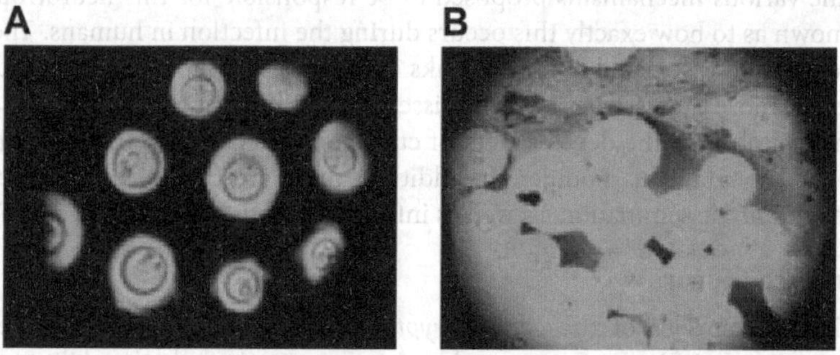

Figure 1. *Cyptococcus* spp. cells with capsule from (A) cutaneous lesion and (B) sputum identified by India Ink (40×).

The culturing of *Cryptococcus* spp. serves to corroborate the diagnosis of the disease (Figure 2). The fungus grows well in various culture media that do not contain cycloheximide (blood agar, Sabouraud agar, and brain–heart infusion agar). For cases of meningoencephalitis, the CSF culture must be repeated 7 days after the beginning of treatment and fortnightly thereafter. The cultures may remain positive for a long time, even over the course of treatment.

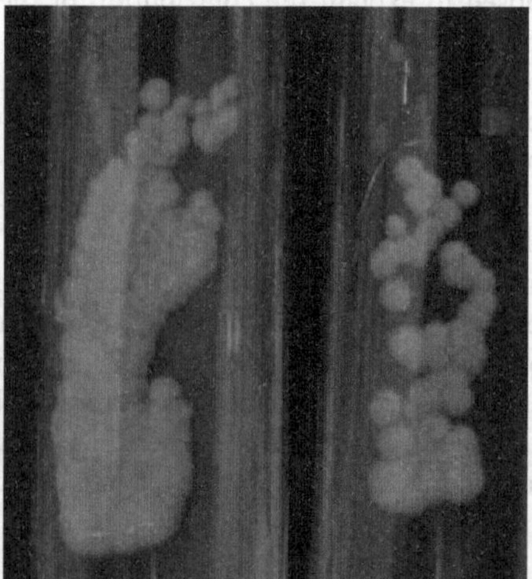

Figure 2. *Cryptococcus* culture in Sabouraud dextrose agar 2%.

After obtaining the isolate, it is necessary to differentiate the species type for clinical and epidemiological purposes [50–52]. Only *C. gattii* is resistant to canavanine and uses glycine as a carbon source; thus, canavanine-glycine-bromothymol blue agar (CGB agar) was proposed in 1982 and has been widely used in laboratories for distinguishing between *C. gattii* and *C. neoformans* species [16, 53, 54]. However, a positive reaction in CGB agar is not sufficient for definitively distinguishing the species as there have been reports of *C. neoformans* resistance to high concentrations of canavanine [55, 56].

The production of urease is a biochemical test used to identify only the genus as both *C. gattii* and *C. neoformans* are able to carry out hydrolysis of urea [57, 58]. The species *C. neofor- mans* and *C. gattii* are the only members of the genus that may produce melanin, thus showing a brownish color in culture media with seed extracts, e.g., *Vicia faba* or *Guizotia abyssinica*. This is due to the presence of tyrosine and chlorogenic acid in these seeds, which are oxidized by phenoloxidase produced by yeast [59, 60]. Culture media that induce the production of melanin are widely used in laboratories for the identification of yeast and differentiation of *Candida* spp.

During infection, the capsular polysaccharides of *C. neoformans* and *C. gattii* solubilize in the body fluids and can be detected and quantified with antibody-based assays. Detection of CrAg (cryptococcal antigen) has been an effective tool for diagnosing cryptococcosis [61, 62]. Detection of CrAg in the serum and CSF by the latex agglutination test (CrAg-latex) or enzyme immunoassays (EIA) has been used for more than 35 years [62]. The majority of comparative studies use cultures as the gold standard. In individuals diagnosed with AIDS and meningitis, the CrAg must always be evaluated, including cases in which the India ink assay cannot identify the yeasts. One study showed that patients living in Uganda with cryptococcal meningitis and HIV who had a negative yeast screening by the India ink assay tested positive for CrAg detection by CSF [63].

The detection of the capsular antigen by agglutination of sensitized particles of latex (LA), which until sometime ago was the immunological method with the most widespread clinical use, may be accomplished in samples from the serum, urine, bronchoalveolar lavage, and CSF. The serological reaction to latex agglutination (LA) is sensitive and specific, emphasizing titers equal to or higher than 1/8 and being able to present cross reaction with the serum of patients with rheumatoid arthritis [62, 64]. The enzyme-linked immunosorbent assay (ELISA) may detect antigens from a cryptococcal infection earlier and at lower titers; however, it is time consuming, expensive, and is laborious. Although CrAg-latex performs as well as EIA and culture, its major limitations are that latex is a manual test and that the resulting interpretation of it is subjective. CrAg-latex and EIA also require laboratory equipment and refrigeration of reagents, making them inadequate for use in environments with minimal infrastructure [61]. The need for refrigeration drastically increases the cost of the test in places with limited resources. Studies report that serological tests with CrAg-latex and EIA may show lower sensitivity when used with strands of some genotypes of *C. gattii*. The rate of false-positive examinations is lower than 1%; false positives are generally explained by technical issues, existence of other infections (e.g., *Trichosporon beigelii*, *Capnocytophaga canimorsus*, and *Stomatococcus mucilaginosus*), or contamination. False-negative results may occasionally be observed with early infections when there is a low fungal charge, with prozone phenomena, and with poorly encapsulated organisms [62, 65].

Recently, a new sensitive, low-cost, fast, and non-laborious immunochromatographic assay known as the lateral flow immunoassay (LFA) was made available for purchase for use in serum, CSF, and urine [66]. This method has demonstrated good sensitivity for the detection of cryptococcal antigen (CrAg), primarily in HIV-positive patients [27]. The World Health Organization (WHO) has recommended the use of antigen detection using LFA for patients infected with HIV who show low CD4 cells and are asymptomatic from a neurological viewpoint [67]. This strategy enables the early identification of patients with a cryptococcal disease in the subclinical stage [68]. It has been used in various studies as a form of screening and diagnosis, thus easing its application to clinical practice. Nevertheless, reasonably good results have been accomplished in multiple types of biological specimens, e.g., blood, CSF, and urine [69].

6. Synthetic peptides

The concept of synthetic peptides and protocols for their artificial synthesis was introduced in the early 20th century [70]. Since then, peptides have become increasingly important for biochemistry, medicine, and biotechnology. In 1963, Bruce Merrifield described the develop- ment of solid-phase peptide synthesis, a technique that made the large-scale production of synthetic peptides a reality. Since then, various studies with different sizes of synthetic peptides have been reported [71].

In the early 1990s, with the development of biotechnology, recombinant antigens were widely used in clinical diagnosis to detect specific antibodies. However, their use in diagnostic tests presented some problems, such as low immunoreactivity compared with the corresponding purified human antigens, laborious and expensive production, and variation in inter-assay reactivity [72–76].

In this regard, synthetic peptides have opened up a new field and perspective as a source of pure epitopes and molecules for the diagnosis of various infectious and noninfectious diseases based on the detection of circulating antibodies and antigens and can also be used for the development of vaccines [77]. Bioinformatics tools are widely used to predict antigenic and immunogenic regions. These programs are capable of predicting B and T cells epitopes, primarily by building on the known properties of amino acids, e.g., their hydrophilicity, charge, flexibility, exposed surface area, and secondary structure [78–80].

Some factors must be taken into account when dealing with synthetic peptides. The first factor to observe is whether the epitopic area is continuous or discontinuous because the amino acids belonging to the epitope are often separated in the linear sequence and become juxtaposed only when the antigen is in its native conformation. The second factor for observation is the size of the epitope. When this field of study began, researchers worked with only small epitopes as prior to the development of solid-phase peptide synthesis, one could not synthesize very large peptides. The very large peptides (>25–30 amino acids) are more expensive and difficult to produce and also have lower yields. For these reasons, peptides of 10–15 amino acid residues are usually recommended for the production and detection of antibodies [81–83].

The use of synthetic peptides for diagnostic tests confers several advantages, e.g., they are innocuous, easy to store and transport, have a high level of reproducibility with low levels of nonspecific reactions, and retain the possibility of changing the chemistry of the peptide by inserting cysteine residues, fatty acids, or carrier proteins or even by incorporating post- translational modifications, such as phosphorylation [84–86].

Over the past 20 years, several peptide sequences have been used to improve the sensitivity and specificity of tests that use recombinant or native protein as antigens [87–93]. However, the use of synthetic peptides as antigens has grown, with many diagnostic systems that are based on synthetic peptides in production, with some being commercially available at the present time. Some diagnostic tests that use synthetic peptides may already be part of the routine clinical diagnosis of certain diseases that involve viruses, parasites, or autoimmune diseases.

Some of the tests that are already available on the market include tests for Epstein–Barr virus, which examines various epitopes on the capsid protein; hepatitis C virus, which includes synthetic peptides that mimic its structural and nonstructural regions (NS4 and NS5); coro- navirus, which is composed of synthetic peptides derived from epitopes of the nucleocapsid and spike proteins and can detect the presence of antibodies from human serum and plasma specimens; *Chlamydia trachomatis*, which has three ELISA diagnostic tests available on the market; and rheumatoid arthritis, with three generations of diagnostic tests based on the detection of antibodies by synthetic peptides [94–97]. Despite the existence of various diag- nostic tests using synthetic peptides and the prevalence of studies reporting the use of synthetic peptides for the diagnosis of various pathologies, particularly those of medical importance, such as tuberculosis, Chagas disease, and leishmaniasis, little has evolved in this area with respect to systemic mycoses [98–103].

Recent advances have been made in the search for more easily available immunodiagnostic tests for fungal infections. Various methods with high specificity and sensitivity are still under development, with a particular emphasis on the search for markers that are able to detect infections at an early stage. In this regard, Caldini et al. [100] used synthetic peptides from the gp75 *Paracoccidioides brasiliensis* antigen as an alternative diagnostic method for the detection of paracoccidioidomycosis.

With regards to cryptococcosis, the search for new diagnostic tests is necessary due to the high incidence of the disease, with estimates of approximately 1 million cases of cryptococcal meningitis per year and more than 600,000 deaths in HIV-infected patients, the potential for *C. gattii* to cause the disease in immunocompetent individuals, and its rapid worldwide dissemination [8]. Therefore, controlling the disease is dependent on epidemiological control of endemic areas coupled with the mapping of new cases and early diagnosis of the disease in affected individuals.

As previously mentioned, diagnostic methods based upon the detection of antibodies have been developed and successfully applied to various other infectious diseases. The efficacy of these methods is not impacted by the antigenic charge of the microorganism, which is particularly relevant for the diagnosis of cryptococcosis, whose major diagnostic tests, LA and LFA, are dependent on the charge of the antigen.

The early diagnosis of cryptococcosis is a challenge that science and the health system must face as in most cases, the disease is diagnosed late, which results in significant morbidity and mortality. Thus, efforts should be made toward finding a rapid, sensitive, and specific diagnosis. In this sense, the identification of multiple immunogenic targets and the possibility of synthesizing these artificial targets appear to be a promising alternative for the development of more accurate tests for the diagnosis of systemic mycosis.

In this area, Martins et al. [104] have adopted an innovative strategy that combines the technology of proteomics and bioinformatics, with the aim of identifying multiple immuno- genic targets for a diagnostic test for cryptococcosis. Linear B-cell epitopes of immunoreactive proteins for *Cryptococcus* species were mapped using *in silico* analyses.

In the search for a faster and more specific test, Brandão et al. [105] tested various synthetic peptides derived from immunoreactive proteins of *Cryptococcus*spp. Of these, six showed good results, which became promising candidate antigens for future diagnostic tests. These six peptides belonged to the proteins Hsp70, Sks2, GrpE, enolase, and two hypothetical proteins. One of these, derived from Hsp70, showed 100% specificity and approximately 80% sensitivity. Table 1 shows the specificity and sensitivity of diverse diagnostic methods for cryptococcosis.

Table 1. Comparative performance of cryptococcosis diagnostic tests.

Tests	Sensitivity (%)	Specificity (%)	References
India ink	30–80	100	[46–48]
Culture	80	100	[106]
CrAg-LA	93–100	93–98	[62]
CrAg-EIA	93–100	93–98	[62]
CrAg-LFA	99–100	92–100	[106]
Synthetic peptides	55–79	90–100	[105]

CrAg, cryptococcal antigen; EIA, enzyme immunoassay; LFA, lateral flow assay; LA, latex agglutination

Hsp70 is a conserved protein that has been increasingly studied worldwide for its role in various biological processes, including the interaction of *Cryptococcus*spp with host cells. Hsp proteins have been characterized as dominant antigens in diverse models, including candi- diasis, aspergillosis, and histoplasmosis [107–110]. In cryptococcosis, the Hsp proteins have been reported to be key antigens that are important for inducing the humoral response [104, 111–113]. These reports support the proposed use of epitopes from immunoreactive proteins, e.g., Hsp70, as antigens in a diagnostic test for cryptococcosis. Higher diagnostic performance can be achieved with multi-epitope chimeric proteins. This type of antigen becomes more attractive because it has more than one antigen-binding site, thus multiplying the possibilities for increasing antigenicity. Brandão et al. demonstrated in a theoretical model (*in silico*) that the combination of peptides in a single molecule is a good strategy for improving the accuracy of a test; therefore, its use is of interest for the development of new diagnostic tests [105, 114–118].

The use of this technology for the development of a diagnostic test capable of the early identification of cryptococcosis and the possibility of building an effective vaccine for this disease are essential for significant reduction in the morbidity and mortality of individuals affected by this infection and may ultimately change their prognosis.

References

1. Del Poeta M, Casadevall A. Ten challenges on Cryptococcus and Cryptococcosis. Mycopathologia. 2012;173(5):303–310. DOI: 10.1007/s11046-011-9473-z
2. Springer DJ, Chaturvedi V. Projecting global occurrence of Cryptococcus gattii. Emerg Infect Dis. 2010;16(1):14–20. DOI: 10.3201/eid1601.090369
3. Byrnes EJ 3rd, Li W, Lewit Y, Ma H, Voelz K, Ren P, Carter DA, Chaturvedi V, Bild- fell RJ, May RC, Heitman J. Emergence and pathogenicity of highly virulent Crypto- coccus gattii genotypes in the northwest United States. PLoS Pathog. 2010;4(1):e1000850. DOI: 10.1371/journal.ppat.1000850
4. D'Souza CA, Kronstad JW, Taylor G, Warren R, Yuen M, Hu G, Jung WH, Sham A, Kidd SE, Tangen K, Lee N, Zeilmaker T, Sawkins J, McVicker G, Shah S, Gnerre S, Griggs A, Zeng Q, Bartlett K, Li W, Wang X, Heitman J, Stajich JE, Fraser JA, Meyer W, Carter D, Schein J, Krzywinski M, Kwon-Chung KJ, Varma A, Wang J, Brunham R, Fyfe M, Ouellette BF, Siddiqui A, Marra M, Jones S, Holt R, Birren BW, Galagan JE, Cuomo CA. Genome variation in Cryptococcus gattii, an emerging pathogen of immunocompetent hosts. MBio. 2011;2(1):e00342-10. DOI: 10.1128/mBio.00342-10
5. Kidd SE, Hagen F, Tscharke RL, Huynh M, Bartlett KH, Fyfe M, Macdougall L, Boekhout T, Kwon-Chung KJ, Meyer W. A rare genotype of Cryptococcus gattii caused the cryptococcosis outbreak on Vancouver Island (British Columbia, Canada). Proc Natl Acad Sci U S A. 2004;101(49):17258-17263.
6. Center of Disease Control and Prevention. Emergence of Cryptococcus gattii – Pacific Northwest, 2004–2010. Am J Transplant. 2011;11(9):1989–1992. DOI: 10.1111/j.1600-6143.2011.03749.x
7. Chen SCA, Meyer W, Sorrell TC. Cryptococcus gattii infections. Clin Microbiol Rev. 2014;27(4):980–1024. DOI: 10.1128/CMR.00126-13
8. Park BJ, Wannemuehler KA, Marston BJ, Govender N, Pappas PG, Chiller TM. Estimation of the current global burden of cryptococcal meningitis among persons living with HIV/AIDS. AIDS. 2009;23(4):525–530. DOI: 10.1097/QAD.0b013e328322ffac
9. Jarvis JN, Lawn SD, Wood R, Harrison TS. Cryptococcal antigen screening for patients initiating antiretroviral therapy: Time for action. Clin Infect Dis. 2010;51(12):1463–1465. DOI: 10.1086/657405
10. Rajasingham R, Rhein J, Klammer K, Musubire A, Nabeta H, Akampurira A, Mossel EC, Williams DA, Boxrud DJ, Crabtree MB, Miller BR, Rolfes MA, Tengsupakul S, Andama AO, Meya DB, Boulware DR. Epidemiology of meningitis in an HIV-infected Ugandan cohort. Am J Trop Med Hyg. 2015;92(2):274–279. DOI: 10.4269/ajtmh.14-0452
11. Vidal JE, Boulware DR. Lateral flow assay for Cryptococcal antigen: An important advance to improve the continuum of HIV care and reduce Cryptococcal meningitis-related mortality. Rev Inst Med Trop Sao Paulo. 2015;57(Suppl 19):38–45. DOI: 10.1590/S0036-46652015000700008
12. Kwon-Chung KJ, Bennett JE. Cryptococcosis. In: Kwon-Chung KJ, Bennett JE, editors. Medical Mycology. Philadelphia: Lea & Febiger; 1992. p. 397–446.

13. Kwon-Chung KJ, Boekhout T, Wickes BL, Fell JW. Systematics of the genus Cryptococcus and its type species C. neoformans. In: Heitman J, Kozel TR, Kwon-Chung KJ, Perfect JR, Casadevall A, editors. Cryptococcus: From human pathogen to model yeast. Washington: ASM Press; 2011. p. 3–15.

14. Bovers M, Hagen F, Boekhout T. Diversity of the Cryptococcus neoformans-Cryptococcus gattii species complex. Rev Iberoam Micol. 2008;25(1):S4–12.

15. Kwon-Chung KJ, Boekhout T, Fell JW, Diaz M. Proposal to conserve the name Cryptococcus gattii against C. hondurianus. Taxon. 2002;51:804–806.

16. Abegg MA, Cella FL, Faganello J, Valente P, Schrank A, Vainstein MH. Cryptococcus neoformans and Cryptococcus gattii isolated from the excreta of psittaciformes in a southern Brazilian zoological garden. Mycopathologia. 2006;161(2):83–91.

17. Boekhout T, Theelen B, Diaz M, Fell JW, Hop WC, Abeln EC, Dromer F, Meyer W. Hybrid genotypes in the pathogenic yeast Cryptococcus neoformans. Microbiology. 2001;147(Pt 4):891–907.

18. Litvintseva AP, Kestenbaum L, Vilgalys R, Mitchell TG. Comparative analysis of environmental and clinical populations of Cryptococcus neoformans. J Clin Microbiol. 2005;43(2):556–564. DOI: 10.1128/JCM.43.2.556-564.2005

19. Meyer W, Castañeda A, Jackson S, Huynh M, Castañeda E; IberoAmerican Cryptococcal Study Group. Molecular typing of IberoAmerican Cryptococcus neoformans isolates. Emerg Infect Dis. 2003;9(2):189–195. DOI: 10.3201/eid0902.020246

20. Ngamskulrungroj P, Gilgado F, Faganello J, Litvintseva AP, Leal AL, Tsui KM, Mitchell TG, Vainstein MH, Meyer W. Genetic diversity of the Cryptococcus species complex suggests that Cryptococcus gattii deserves to have varieties. PLoS One. 2009;4(6):e5862. DOI: 0.1371/journal.pone.0005862

21. MacDougall L, Kidd SE, Galanis E, Mak S, Leslie MJ, Cieslak PR, Kronstad JW, Morshed MG, Bartlett KH. Spread of Cryptococcus gattii in British Columbia, Canada, and detection in the Pacific Northwest, USA. Emerg Infect Dis. 2007;13(1):42–50. DOI: 10.3201/eid1301.060827

22. Datta K, Bartlett KH, Baer R, Byrnes E, Galanis E, Heitman J, Hoang L, Leslie MJ, MacDougall L, Magill SS, Morshed MG, Marr KA; Cryptococcus gattii Working Group of the Pacific Northwest. Spread of Cryptococcus gattii into Pacific Northwest region of the United States. Emerg Infect Dis. 2009;15(8):1185–1191. DOI: 10.3201/eid1508.081384

23. Trilles L, Lazéra M, Wanke B, Theelen B, Boekhout T. Genetic characterization of en- vironmental isolates of the Cryptococcus neoformans species complex from Brazil. Med Mycol. 2003;41(5):383–390. DOI: 10.1080/1369378031000137206

24. Trilles L, Lazéra Mdos S, Wanke B, Oliveira RV, Barbosa GG, Nishikawa MM, Morales BP, Meyer W. Regional pattern of the molecular types of Cryptococcus neoformans and Cryptococcus gattii in Brazil. Mem Inst Oswaldo Cruz. 2008;103(5):455–462. DOI: 10.1590/S0074-02762008000500008

25. Chen S, Sorrell T, Nimmo G, Speed B, Currie B, Ellis D, Marriott D, Pfeiffer T, Parr D, Byth K. Epidemiology and host- and variety-dependent characteristics of infection

due to Cryptococcus neoformans in Australia and New Zealand. Clin Infect Dis. 2000;31(2):499–508. DOI: 10.1086/313992

26. Ellis D, Marriott D, Hajjeh RA, Warnock D, Meyer W, Barton R. Epidemiology: Sur- veillance of fungal infections. Med Mycol. 2000;38(Suppl 1):173–182.

27. Harris J, Lockhart S, Chiller T. Cryptococcus gattii: Where do we go from here?. Med Mycol. 2012;50(2):113–129. DOI: 10.3109/13693786.2011.607854

28. Choi YH, Ngamskulrungroj P, Varma A, Sionov E, Hwang SM, Carriconde F, Meyer W, Litvintseva AP, Lee WG, Shin JH, Kim EC, Lee KW, Choi TY, Lee YS, Kwon-Chung KJ. Prevalence of the VNIc genotype of Cryptococcus neoformans in non- HIV-associated cryptococcosis in the Republic of Korea. FEMS Yeast Res. 2010;10(6): 769–778. DOI: 10.1111/j.1567-1364.2010.00648.x

29. Viviani MA, Cogliati M, Esposto MC, Lemmer K, Tintelnot K, Colom Valiente MF, Swinne D, Velegraki A, Velho R; European Confederation of Medical Mycology (ECMM) Cryptococcosis Working Group. Molecular analysis of 311 Cryptococcus neoformans isolates from a 30-month ECMM survey of cryptococcosis in Europe. FEMS Yeast Res. 2006;6(4):614–619. DOI: 10.1111/j.1567-1364.2006.00081.x

30. Li SS, Mody CH. Cryptococcus. Proc Am Thorac Soc. 2010;7(3):186–196. DOI: 10.1513/pats.200907-063AL

31. Jongwutiwes U, Sungkanuparph S, Kiertiburanakul S. Comparison of clinical fea- tures and survival between cryptococcosis in human immunodeficiency virus (HIV)- positive and HIV-negative patients. Jpn J Infect Dis. 2008;61(2):111–115.

32. Moretti ML, Resende MR, Lazéra MS, Colombo AL, Shikanai-Yasuda MA. Guide-lines in cryptococcosis – 2008. Rev Soc Bras Med Trop. 2008;41(5):524–544.

33. Cohen J, Perfect JR, Durack DT. Cryptococcosis and the basidiospore. Lancet. 1982;319(8284):1301. DOI: 10.1016/S0140-6736(82)92861-6

34. Lin X, Heitman J. The biology of the Cryptococcus neoformans species complex. Annu Rev Microbiol. 2006;60:69–105. DOI: 10.1146/annurev.micro.60.080805.142102

35. Littman ML, Schneierson SS. Cryptococcus neoformans in pigeon excreta in New York City. Am J Hyg. 1959;69(1):49–59.

36. Krockenberger MB, Malik R, Ngamskulrungroj P, Trilles L, Escandon P, Dowd S, Allen C, Himmelreich U, Canfield PJ, Sorrell TC, Meyer W. Pathogenesis of pulmonary Cryptococcus gattii infection: A rat model. Mycopathologia. 2010;170(5):315–330. DOI: 10.1007/s11046-010-9328-z. Epub 2010 Jun 15

37. Weatherhead SC, Charlton FG, Reynolds NJ. Plaques, papules, and nodules in a 40-Year-Old Man. Arch Dermatol. 2006;142(7):921–926. DOI: doi:10.1001/archderm.142.7.921-a

38. Kronstad JW, Attarian R, Cadieux B, Choi J, D'Souza CA, Griffiths EJ, Geddes JM, Hu G, Jung WH, Kretschmer M, Saikia S, Wang J. Expanding fungal pathogenesis: Cryptococcus breaks out of the opportunistic box. Nat Rev Microbiol. 2011;9(3):193–203. DOI: 10.1038/nrmicro2522

39. Kwon-Chung KJ, Fraser JA, Doering TL, Wang Z, Janbon G, Idnurm A, Bahn YS.

Cryptococcus neoformans and Cryptococcus gattii, the etiologic agents of cryptococcosis. Cold Spring Harb Perspect Med. 2014;4(7):a019760. DOI: 10.1101/cshperspect.a019760

40. McFadden DC, De Jesus M, Casadevall A. The physical properties of the capsular polysaccharides from Cryptococcus neoformans suggest features for capsule construction. J Biol Chem. 2006;281(4):1868–1875. DOI: 10.1074/jbc.M509465200

41. Rakesh V, Schweitzer AD, Zaragoza O, Bryan R, Wong K, Datta A, Casadevall A, Da- dachova E. Finite-element model of interaction between fungal polysaccharide and monoclonal antibody in the capsule of Cryptococcus neoformans. J Phys Chem B. 2008;112(29):8514–8522. DOI: 10.1021/jp8018205

42. Bose I, Reese AJ, Ory JJ, Janbon G, Doering TL. A yeast under cover: The capsule of Cryptococcus neoformans. Eukaryot Cell. 2003;2(4):655–663. DOI: 10.1128/EC.2.4.655-663.2003

43. Collins HL, Bancroft GJ. Encapsulation of Cryptococcus neoformans impairs antigen-specific T-cell responses. Infect Immun. 1991;59(11):3883–3888.

44. Monari C, Paganelli F, Bistoni F, Kozel TR, Vecchiarelli A. Capsular polysaccharide induction of apoptosis by intrinsic and extrinsic mechanisms. Cell Microbiol. 2008;10(10):2129–2137. DOI: 10.1111/j.1462-5822.2008.01196.x

45. Beyene T, Woldeamanuel Y, Asrat D, Ayana G, Boulware DR. Comparison of cryptococcal antigenemia between antiretroviral naïve and antiretroviral experienced HIV positive patients at two hospitals in Ethiopia. PLoS One. 2013;8(10):e75585. DOI: 10.1371/journal.pone.0075585

46. Saha DC, Xess I, Jain N. Evaluation of conventional & serological methods for rapid diagnosis of cryptococcosis. Indian J Med Res. 2008;127(5):483–488.

47. Snow RM, Dismukes WE. Cryptococcal meningitis: Diagnostic value of cryptococcal antigen in cerebrospinal fluid. Arch Intern Med. 1975;135(9):1155–1157.

48. Imwidthaya P, Poungvarin N. Cryptococcosis in AIDS. Postgrad Med J. 2000;76:85–88.

49. Namiq AL, Tollefson T, Fan F. Cryptococcal parotitis presenting as a cystic parotid mass: Report of a case diagnosed by fine-needle aspiration cytology. Diagn Cytopathol. 2005;33(1):36–38.

50. Fries BC, Goldman DL, Cherniak R, Ju R, Casadevall A. Phenotypic switching in Cryptococcus neoformans results in changes in cellular morphology and glucuronoxylomannan structure. Infect Immun. 1999;67(11):6076–6083.

51. Fries BC, Taborda CP, Serfass E, Casadevall A. Phenotypic switching of Cryptococcus neoformans occurs in vivo and influences the outcome of infection. J Clin Invest. 2001;108(11):1639–1648.

52. Lacaz CS, Porto E, Martins JEC, Heins-Vaccari EM. Criptococose. In: Lacaz CS, Porto E, Martins JEC, Heins-Vaccari EM, Melo NT, editors. Tratado de micologia, 1st ed. São Paulo: Sarvier; 2002. p. 416–440.

53. Canelo C, Navarro A, Guevara M, Urcia F, Zurita S, Casquero J. Determinación de

la variedad de cepas de Cryptococcus neoformasn aisladas de pacientes con SIDA. Rev Med Exp. 1999;15(1–2):44–47.

54. Kwon-Chung KJ, Polacheck I, Popkin TJ. Melanin-lacking mutants of Cryptococcus neoformans and their virulence for mice. J Bacteriol. 1982;150(3):1414–1421.

55. Khan ZU, Al-Anezi AA, Chandy R, Xu J. Disseminated cryptococcosis in an AIDS patient caused by a canavanine-resistant strain of Cryptococcus neoformans var. gru- bii. J Med Microbiol. 2003;52(Pt 3):271–275. DOI: 10.1099/jmm.0.05097-0

56. Nakamura Y, Kano R, Sato H, Watanabe S, Takahashi H, Hasegawa A. Isolates of Cryptococcus neoformans serotype A and D developed on canavanine-glycine-brom- thymol blue medium. Mycoses. 1998;41(1–2):35–40. DOI: 10.1111/j. 1439-0507.1998.tb00373.x

57. Cox GM, Mukherjee J, Cole GT, Casadevall A, Perfect JR. Urease as a virulence factor in experimental cryptococcosis. Infect Immun. 2000;68(2):443–448. DOI: 10.1128/IAI. 68.2.443-448.2000

58. Casali AK, Goulart L, Rosa e Silva LK, Ribeiro AM, Amaral AA, Alves SH, Schrank A, Meyer W, Vainstein MH. Molecular typing of clinical and environmental Cryptococcus neoformans isolates in the Brazilian state Rio Grande do Sul. FEMS Yeast Res. 2003;3(4):405–415. DOI: 10.1016/S1567-1356(03)00038-2

59. Denning DW, Stevens DA, Hamilton JR. Comparison of Guizotia abyssinica seed ex- tract (birdseed) agar with conventional media for selective identification of Crypto- coccus neoformans in patients with acquired immunodeficiency syndrome. J Clin Microbiol. 1990;28(11):2565–2567.

60. Jacobson ES. Pathogenic roles for fungal melanins. Clin Microbiol Rev. 2000;13(4): 708–717.

61. Makadzange AT, McHugh G. New approaches to the diagnosis and treatment of cryptococcal meningitis. Semin Neurol. 2014;34(1):47–60. DOI: 10.1055/ s-0034-1372342

62. Perfect JR, Bicanic T. Cryptococcosis diagnosis and treatment: What do we know now. Fungal Genet Biol. 2015;78:49–54. DOI: 10.1016/j.fgb.2014.10.003

63. Boulware DR, Rolfes MA, Rajasingham R, von Hohenberg M, Qin Z, Taseera K, Schutz C, Kwizera R, Butler EK, Meintjes G, Muzoora C, Bischof JC, Meya DB. Multisite validation of cryptococcal antigen lateral flow assay and quantification by laser thermal contrast. Emerg Infect Dis. 2014;20(1):45–53. DOI: 10.3201/eid2001.130906

64. Mitchell TG, Perfect JR. Crytococcosis in the era of AIDS – 100 years after the discovery of Cryptococcus neoformans. Clin Microbiol Rev. 1995;8(4):515–548.

65. Antinori S. New insights into HIV/AIDS-associated cryptococcosis. ISRN AIDS. 2013:471363. DOI: 10.1155/2013/471363

66. Lindsley MD, Mekha N, Baggett HC, Surinthong Y, Autthateinchai R, Sawatwong P, Harris JR, Park BJ, Chiller T, Balajee SA, Poonwan N. Evaluation of a newly developed lateral flow immunoassay for the diagnosis of cryptococcosis. Clin Infect Dis. 2011;53(4):321–325. DOI: 10.1093/cid/cir379

67. Rajasingham R, Meya DB, Boulware DR. Integrating cryptococcal antigen screening

and pre-emptive treatment into routine HIV care. J Acquir Immune Defic Syndr. 2012;59(5):e85–91. DOI: 10.1097/QAI.0b013e31824c837e

68. World Health Organization. Rapid Advice: Diagnosis, Prevention and Management of Cryptococcal Disease in HIV-Infected Adults, Adolescents and Children. Geneva: World Health Organization; 2011. 44. DOI: 13: 978-92-4-150297-9

69. Rugemalila J, Maro VP, Kapanda G, Ndaro AJ, Jarvis JN. Cryptococcal antigen prev- alence in HIV-infected Tanzanians: A cross-sectional study and evaluation of a point- of-care lateral flow assay. Trop Med Int Health. 2013;18(9):1075–1079. DOI: 10.1111/ tmi.12157.

70. Wieland T, Bodanszky M. The World of Peptides: A Brief History of Peptide Chemistry. 1st ed. Berlin: Springer-Verlag; 1991. 298. DOI: 10.1007/978-3-642-75850-8

71. Merrifield RB. Solid phase peptide synthesis. I. The synthesis of a tetrapeptide. J Am Chem Soc. 1963;85(14):2149–2154. DOI: 10.1021/ja00897a025

72. Gonzalez L, Boyle RW, Zhang M, Castillo J, Whittier S, Della-Latta P, Clarke LM, George JR, Fang X, Wang JG, Hosein B, Wang CY. Synthetic-peptide-based enzyme-linked immunosorbent assay for screening human serum or plasma for antibodies to human immunodeficiency virus type 1 and type 2. Clin Diagn Lab Immunol. 1997;4(5):598–603.

73. Leinikki P, Lehtinen M, Hyöty H, Parkkonen P, Kantanen ML, Hakulinen J. Synthetic peptides as diagnostic tools in virology. Adv Virus Res. 1993;42:149–186.

74. Routsias JG, Tzioufas AG, Moutsopoulos HM. The clinical value of intracellular auto- antigens B-cell epitopes in systemic rheumatic diseases. Clin Chim Acta. 2004;340(1– 2):1–25. DOI: 10.1016/j.cccn.2003.10.011

75. Thorpe R, Wadhwa M, Mire-Sluis A. The use of bioassays for the characterisation and control of biological therapeutic products produced by biotechnology. Dev Biol Stand. 1997;91:79–88.

76. Yan SC, Grinnell BW, Wold F. Post-translational modifications of proteins: Some problems left to solve. Trends Biochem Sci. 1989;14(7):264–268. DOI: 10.1016/0968-0004(89)90060-1

77. Noya O, Patarroyo ME, Guzmán F, Alarcón de Noya B. Immunodiagnosis of parasitic diseases with synthetic peptides. Curr Protein Pept Sci. 2003;4(4):299–308. DOI: 10.2174/1389203033487153

78. Sun P, Ju H, Liu Z, Ning Q, Zhang J, Zhao X, Huang Y, Ma Z, Li Y. Bioinformatics resources and tools for conformational B-cell epitope prediction. Comput Math Methods Med. 2013;2013:943636. DOI: 10.1155/2013/943636

79. Greenbaum JA, Andersen PH, Blythe M, Bui HH, Cachau RE, Crowe J, Davies M, Kolaskar AS7, Lund O, Morrison S, Mumey B, Ofran Y, Pellequer JL, Pinilla C, Pono- marenko JV, Raghava GPS, Van Regenmortel MHV, Roggen EL, Sette A, Schlessinger A, Sollner J, Zand M, Peters B. Towards a consensus on datasets and evaluation met- rics for developing B-cell epitope prediction tools. J Mol Recognit. 2007;20:75–82. DOI: 10.1002/jmr.815

80. Yao B, Zheng D, Liang S, Zhang C. Conformational B-cell epitope prediction on

anti- gen protein structures: A review of current algorithms and comparison with com- mon binding site prediction methods. PLoS One. 2013;8(4):e62249. DOI: 10.1371/ journal.pone.0062249

81. Greenfield EA, DeCaprio J, Brahmandam M. Selecting the antigen. In: Greenfield EA, editor. Antibodies: A laboratory manual. 2nd ed. New York: Cold Spring Harbor Press; 2013. p. 43–106.

82. Moore ML, Grant GA. Peptide design considerations. In: Grant GA, editor. Synthetic peptide: A user's guide. 2nd ed. New York: Oxford; 2002. p. 10–92.

83. Davies DH, Halablab MA, Clarker J, Cox FEG, Young TWK. Infection and Immunity. 1st ed. London: Taylor & Francis; 2003. 237.

84. Gómara MJ, Haro I. Synthetic peptides for the immunodiagnosis of human diseases. Curr Med Chem. 2007;14(5):531–546.

85. Wu CL, Leu TS, Chang TT, Shiau AL. Hepatitis C virus core protein fused to hepatitis B virus core antigen for serological diagnosis of both hepatitis C and hepatitis B infections by ELISA. J Med Virol. 1999;57(2):104–110.

86. Hans D, Young PR, Fairlie DP. Current status of short synthetic peptides as vaccines. Med Chem. 2006;2(6):627–646.

87. El Awady MK, El-Demellawy MA, Khalil SB, Galal D, Goueli SA. Synthetic peptide-based immunoassay as a supplemental test for HCV infection. Clin Chim Acta. 2002;325(1–2):39–46. DOI: 10.1016/S0009-8981(02)00245-0

88. Favorov MO, Khudyakov YE, Fields HA, Khudyakova NS, Padhye N, Alter MJ, Mast E, Polish L, Yashina TL, Yarasheva DM, Onischenkob GG, Margolis HS. Enzyme im- munoassay for the detection of antibody to hepatitis E virus based on synthetic pep- tides. J Virol Methods. 1994;46(2):237–250.

89. Gevorkian G, Soler C, Viveros M, Padilla A, Govezensky T, Larralde C. Serologic re- activity of a synthetic peptide from human immunodeficiency virus type 1 gp41 with sera from a Mexican population. Clin Diagn Lab Immunol. 1996;3(6):651–653.

90. Gnann JW Jr, McCormick JB, Mitchell S, Nelson JA, Oldstone MB. Synthetic peptide immunoassay distinguishes HIV type 1 and HIV type 2 infections. Science. 1987;237(4820):1346–1349.

91. Hernández M, Beltrán C, García E, Fragoso G, Gevorkian G, Fleury A, Parkhouse M, Harrison L, Sotelo J, Sciutto E. Cysticercosis: Towards the design of a diagnostic kit based on synthetic peptides. Immunol Lett. 2000;71(1):13–17. DOI: 10.1016/ S0165-2478(99)00166-2

92. Mahler M, Blüthner M, Pollard KM. Advances in B-cell epitope analysis of autoantigens in connective tissue diseases. Clin Immunol. 2003;107(2):65–79. DOI: 10.1016/ S1521-6616(03)00037-8

93. Shin SY, Lee MK, Kim SY, Jang SY, Hahm KS. The use of multiple antigenic peptide (MAP) in the immunodiagnosis of human immunodeficiency virus infection. Biochem Mol Biol Int. 1997;43(4):713–721.

94. Chan PK, To WK, Liu EY, Ng TK, Tam JS, Sung JJ, Lacroix JM, Houde M. Evaluation

of a peptide-based enzyme immunoassay for anti-SARS coronavirus IgG antibody. J Med Virol. 2004;74(4):517–520. DOI: 10.1002/jmv.20207

95. Fachiroh J, Paramita DK, Hariwiyanto B, Harijadi A, Dahlia HL, Indrasari SR, Kusumo H, Zeng YS, Schouten T, Mubarika S, Middeldorp JM. Single-assay combination of Epstein-Barr Virus (EBV) EBNA1- and viral capsid antigen-p18-derived synthetic peptides for measuring anti-EBV immunoglobulin G (IgG) and IgA antibody levels in sera from nasopharyngeal carcinoma patients: Options for field screening. J Clin Microbiol. 2006;44(4):1459–1467. DOI: 10.1128/JCM.44.4.1459-1467.2006

96. Schellekens GA, Visser H, de Jong BA, van den Hoogen FH, Hazes JM, Breedveld FC, van Venrooij WJ. The diagnostic properties of rheumatoid arthritis antibodies recognizing a cyclic citrullinated peptide. Arthritis Rheum. 2000;43(1):155–163.

97. Yoshida CF, Rouzerè CD, Nogueira RM, Lampe E, Travassos-da-Rosa MA, Vanderborght BO, Schatzmayr HG. Human antibodies to dengue and yellow fever do not react in diagnostic assays for hepatitis C virus. Braz J Med Biol Res. 1992;25(11):1131–1135.

98. Araujo Z, Giampietro F, Bochichio Mde L, Palacios A, Dinis J, Isern J, Waard JH, Rada E, Borges R, Fernández de Larrea C, Villasmil A, Vanegas M, Enciso-Moreno JA, Patarroyo MA. Immunologic evaluation and validation of methods using synthetic peptides derived from Mycobacterium tuberculosis for the diagnosis of tuberculosis infection. Mem Inst Oswaldo Cruz. 2013;108(2):131–139. DOI: 10.1590/0074-0276108022013001

99. Bottino CG, Gomes LP, Pereira JB, Coura JR, Provance DW Jr, De-Simone SG. Chagas disease-specific antigens: Characterization of epitopes in CRA/FRA by synthetic pep- tide mapping and evaluation by ELISA-peptide assay. BMC Infect Dis. 2013;13:568. DOI: 10.1186/1471-2334-13-568

100. Caldini CP, Xander P, Kioshima ÉS, Bachi AL, de Camargo ZP, Mariano M, Lopes JD. Synthetic peptides mimic gp75 from Paracoccidioides brasiliensis in the diagnosis of paracoccidioidomycosis. Mycopathologia. 2012;174(1):1–10. DOI: 10.1007/s11046-011-9518-3

101. Faria AR, Costa MM, Giusta MS, Grimaldi G Jr, Penido ML, Gazzinelli RT, Andrade HM. High-throughput analysis of synthetic peptides for the immunodiagnosis of ca- nine visceral leishmaniasis. PLoS Negl Trop Dis. 2011;5(9):e1310. DOI: 10.1371/jour- nal.pntd.0001310

102. Gori A, Longhi R, Peri C, Colombo G. Peptides for immunological purposes: Design, strategies and applications. Amino Acids. 2013;45(2):257–268. DOI: 10.1007/s00726-013-1526-9

103. Shen G, Behera D, Bhalla M, Nadas A, Laal S. Peptide-based antibody detection for tuberculosis diagnosis. Clin Vaccine Immunol. 2009;16(1):49-54. DOI: 10.1128/CVI. 00334-08

104. Martins LM, de Andrade HM, Vainstein MH, Wanke B, Schrank A, Balaguez CB, dos Santos PR, Santi L, Pires Sda F, da Silva AS, de Castro JA, Brandão RM, do Monte SJ. Immunoproteomics and immunoinformatics analysis of Cryptococcus gattii:

Novel candidate antigens for diagnosis. Future Microbiol. 2013;8(4):549–563. DOI: 10.2217/ fmb.13.22

105. de Serpa Brandão RM, Soares Martins LM, de Andrade HM, Faria AR, Soares Leal MJ, da Silva AS, Wanke B, dos Santos Lazéra M, Vainstein MH, Mendes RP, Moris DV, de Souza Cavalcante R, do Monte SJ. Immunoreactivity of synthetic peptides de- rived from proteins of Cryptococcus gattii. Future Microbiol. 2014;9(7):871–878. DOI: 10.2217/fmb.14.49

106. McMullan BJ, Halliday C, Sorrell TC, Judd D, Sleiman S, Marriott D, Olma T, Chen SC. Clinical utility of the cryptococcal antigen lateral flow assay in a diagnostic my- cology laboratory. PLoS One. 2012;7(11):e49541. DOI: 10.1371/journal.pone.0049541.

107. Eroles P, Sentandreu M, Elorza MV, Sentandreu R. The highly immunogenic enolase and Hsp70p are adventitious Candida albicans cell wall proteins. Microbiology. 1997;143:313–320. DOI: 10.1099/00221287-143-2-31

108. Burnie JP, Matthews RC. Heat shock protein 88 and Aspergillus infection. J Clin Microbiol. 1991;29(10):2099–2106.

109. Gomez FJ, Gomez AM, Deepe GS Jr. An 80-kilodalton antigen from Histoplasma cap- sulatum that has homology to heat shock protein 70 induces cell-mediated immune responses and protection in mice. Infect Immun. 1992;60(7):2565–2571.

110. Deepe GS Jr, Gibbons RS. Cellular and molecular regulation of vaccination with heat shock protein 60 from Histoplasma capsulatum. Infect Immun. 2002;70(7):3759–3767.

111. Kakeya H, Udono H, Ikuno N, Yamamoto Y, Mitsutake K, Miyazaki T, Tomono K, Koga H, Tashiro T, Nakayama E, Kohno S. A 77-kilodalton protein of Cryptococcus neoformans, a member of the heat shock protein 70 family, is a major antigen detected in the sera of mice with pulmonary cryptococcosis. Infect Immun. 1997;65(5): 1653–1658.

112. Kakeya H, Udono H, Maesaki S, Sasaki E, Kawamura S, Hossain MA, Yamamoto Y, Sawai T, Fukuda M, Mitsutake K, Miyazaki Y, Tomono K, Tashiro T, Nakayama E, Kohno S. Heat shock protein 70 (hsp70) as a major target of the antibody response in patients with pulmonary cryptococcosis. Clin Exp Immunol. 1999;15(3):485–490. DOI: 10.1046/j.1365-2249.1999.00821.x

113. Silveira CP, Piffer AC, Kmetzsch L, Fonseca FL, Soares DA, Staats CC, Rodrigues ML, Schrank A, Vainstein MH. The heat shock protein (Hsp) 70 of Cryptococcus neo- formans is associated with the fungal cell surface and influences the interaction be- tween yeast and host cells. Fungal Genet Biol. 2013;60:53–63. DOI: 10.1016/j. fgb. 2013.08.005

114. Dai J, Jiang M, Wang Y, Qu L, Gong R, Si J. Evaluation of a recombinant multiepitope peptide for serodiagnosis of Toxoplasma gondii infection. Clin Vaccine Immunol. 2012;19(3):338–342. DOI: 10.1128/CVI.05553-11

115. Camussone C, Gonzalez V, Belluzo MS, Pujato N, Ribone ME, Lagier CM, Marcipar IS. Comparison of recombinant Trypanosoma cruzi peptide mixtures versus multiepitope chimeric proteins as sensitizing antigens for immunodiagnosis. Clin Vaccine

Immunol. 2009;16(6):899–905. DOI: 10.1128/CVI.00005-09

116. Faria AR, de Castro Veloso L, Coura-Vital W, Reis AB, Damasceno LM, Gazzinelli RT, Andrade HM. Novel recombinant multiepitope proteins for the diagnosis of asymptomatic leishmania infantum-infected dogs. PLoS Negl Trop Dis. 2015;9(1):e3429. DOI: 10.1371/journal.pntd.0003429

117. Cheng Z, Zhao JW, Sun ZQ, Song YZ, Sun QW, Zhang XY, Zhang XL, Wang HH, Guo XK, Liu YF, Zhang SL. Evaluation of a novel fusion protein antigen for rapid se- rodiagnosis of tuberculosis. J Clin Lab Anal. 2011;25(5):344–349. DOI: 10.1002/jcla. 20483

118. Duthie MS, Hay MN, Morales CZ, Carter L, Mohamath R, Ito L, Oyafuso LK, Manini MI, Balagon MV, Tan EV, Saunderson PR, Reed SG, Carter D. Rational design and evaluation of a multiepitope chimeric fusion protein with the potential for leprosy di- agnosis. Clin Vaccine Immunol. 2010;17(2):298–303. DOI: 10.1128/CVI.00400-09

Chapter-7

Fungal Diseases Occurring on Trees of Namibia

1. Introduction

Fungal plant pathogens, if not controlled one way or the other, can have devastating effects on biodiversity, forest structure and dynamics, commercial plantations, agro forestry and urban environments. This is especially the case with introduced (exotic) pathogens. A well- known example is that of *Cryphonectria parasitica*, a fungus native to Asia, which was intro- duced into the United States in the early 1900s. This fungus, a mild pathogen in its areas of origin, after introducing it to the United States caused the near extinction of North American chestnut trees (*Castanea dentata*). Today, a once-dominant canopy tree in the Eastern United States has been reduced to a low-growing shrub and the entire ecology of the forests has been changed, impacting on animals, other trees and humans. Another example is that of sooty baobab disease, which has killed a lot of baobabs in Southern Africa [1-4] The impact on plantation forestry species and agricultural crops can be equally severe and with the increased movement of humans and plant products around the world, more and more pests and diseases are being moved to areas where they previously did not occur.

Die-back on A*cacia* trees in areas in and around the city of Windhoek in Namibia, especially in the Dorado Park area, was observed and since then, disease symptoms and tree deaths have been increasing in Namibia forest lands [3]. Life-threatening basal cankers on *Syzygium guineense* trees have been observed along the Zambezi River in the Katima Mulilo area of Namibia. *Acacia* species are particularly important for their use as fodder and fuel wood (see Figure 1) for enhancing soil fertility through biological nitrogen fixation and gum production, whereas *Syzygium guineense* is important source of edible fruits and fuel wood. In this study, there are two main objectives: firstly, to investigate the cause of decline and death of *Acacia* species trees around Windhoek, to develop management strategies to reduce the impact of the disease and secondly, to survey fungal pathogens occurring in selected habitats of *Syzygium* and *Acacia* species in Namibia. These surveys will serve as the groundwork for future inves- tigations on tree health in Namibia, that is, in expanding our knowledge pertaining to the nationwide distribution, impact and origin of these pathogens. This knowledge is also essential to understand the functional dynamics of these native ecosystems at a basic level, which is potentially important for the conservation of these ecosystems.

Yet another close example is that of the baobab disease threat. Baobab trees are very resilient and can survive an extraordinary amount of abuse by humans, animals and natural conditions. Very little, however, is known about the diseases that affect these trees. In recent years there have been numerous reports on baobab tree mortality. In most cases the causes of these deaths are not known. The first reports of diseased baobabs are from 1944, when it was reported that baobabs were dying subsequent to being infected with "a browny black smut" after prolonged drought in Zimbabwe [5]. Numerous isolations were attempted from diseased trees but no pathogens were obtained, and it was concluded that the symptoms were a result of environ- mental stresses that made trees susceptible to colonization by sooty mould fungi (Hopkins 1950 as cited by [5]). The same symptoms were noted in Zimbabwe in the 1960s and again in the 1980s [5], as well as in South Africa in the early 1990s [2]. In 2002, a pilot study was conducted to determine the cause of death of baobab trees in the Musina nature reserve [6]. Die-back of branches and exudation of large amounts of sap were noted, but no sooty mould was evident. Internal discoloration of diseased branches was also observed and diseased trees died within a few months.

The death of several large baobabs [5] in the Nyae Nyae conservancy in Namibia: Both sooty mould and the exudation symptoms were observed on the trees. It, therefore, appears that the cause of death may be a combination of diseases or an unidentified pathogen. *Lasiodiplodia theobromae* was isolated from these trees. This fungus belongs to the Botryosphaeriaceae family, in which many of the species included are plant pathogens [7]; however some of the species in this family are endophytes that live in plants and only start to cause disease when the right conditions occur [8].

Figure 1. Ecosystem services from indigenous plants. In panel A: edible fruits of the monkey oranges on the road side in Kavango region of Namibia, panel B: fire wood harvested from local forests in Kavango region being sold on the road side, panel C: bird nests habitat on *Acacia karroo* trees, panel D: pods of *Acacia erioloba* packed in bags for sell on the roadside for feeding livestock, panel E: twig die-back symptoms manifesting on *Acacia karroo*, panel F: *Acacia karroo* tree totally collapsing after fungal infection.

The increasing number of reports of diseased and dying trees in Namibia [9-13] emphasizes the need for urgent investigation of the causes of these deaths. It is necessary to determine whether these reports are isolated incidences or if there is an epidemic underway that can threaten the survival of indigenous trees. Furthermore, it is important to understand the fungal assemblag- es of these trees to understand possible future disease outbreaks. One of the greatest challeng- es to tree health management is the fact that so little is known regarding the fungal biodiversity on trees and the role these fungi play in tree health. In this chapter, the symptoms observed on dying trees and potential causal agents of death are reported. In plant health biology and forestry conservation, the firstdedicated step is to determine causes of plant diseases, which is a primary objective of this investigation. Yearly, calls have been received from farmers about plants dying or manifesting life-threatening symptoms [9-14] (see Figure 1). Some of these reports have been partially investigated and yet others remain to be investigated.

2. Materials and methods

2.1. Description of study site and population

The study was carried out in Oshikoto and Otjozondjupa regions of Namibia, where the disease reports on trees have been received from. In addition, the survey was carried out on forest stands where some of the species occur in Erongo, East and West Kavango and Zambezi regions.

Figure 2. Symptoms observed on selected indigenous trees of Namibia. On panel A: discoloration of the vascular tissue due to fungal infections; panel B: severe malformations of pods on *Acacia karroo*; panel C: malformation of flower in- florescences of *Aloe zebrina*; panel D: *Acacia erioloba* infected with a bracket fungus; panel E: *Welwitschia mirabilis* infect- ed with fungi; panel F: basal canker on *Syzygium guineense* showing sporulation of fungi on the stems along the banks of the Zambezi river in Zambezi region of Namibia.

The study was exploratory in nature and used standard mycological methods to investigate the identity of fungi causing diseases on plant species based on previous report calls received and reports existing in the media. In cases where no reports have been made, field surveys were carried out to observe symptoms in native forest stands.

During the survey, diseased samples from leaves and small branches and twigs were collected (see Figure2). These were then brought to the laboratory at the University of Namibia for isolation of fungal organisms. The pure cultures of these microbial organisms were then stored in a type culture collection facility and used for further classical microbiological analyses and genetic analyses to determine the identity of the organisms. Fungi were identified based on morphology and ITS DNA sequencing data. Further, the pure cultures were used to test Koch's postulates by inoculating the possible pathogens into healthy seedlings.

For the *Welwitschia* and *Syzygium* work, non-destructive sampling was used. Small pieces of infected/cankered tissue were removed for isolation of the fungi. The collection of these pieces of material did not result in the death or deformation of the plants. In the case of the *Welwit- schia*, small pieces, less than 5 cm in length and 1 cm wide, were removed for study. The *Acacia* work was somewhat more destructive because entire sections of dying tree parts were needed to be collected to determine the cause of death and, in the long-term, to prevent other trees from dying. This involved chopping of stems and branches and possibly digging up the roots of affected plants if required.

Survey trips were conducted in various sites in Namibia, including Windhoek, Dordabis, Grootfontein, Katima Mulilo, Omaruru, Swakopmund, Rundu, Popa Falls and Rehoboth between 2003 and 2013 to collect leaves, stems and roots of *Acacia* and *Syzygium* plants showing disease symptoms. At least 10 trees were sampled from each site. Collected samples were examined for insect damage and used for fungal isolations. Primary isolations and pure cultures of fungi were made using 2% MEA media. All isolates were identified in the laboratory using morphology and ribosomal DNA sequencing of the internal transcribed spacer (ITS) regions. The ZR Fungal/Bacterial DNA extraction kit was used to obtain DNA from the crushed powder-form mycelium of the fungi isolated from these trees. ITS 1 and ITS 4 primers were used to obtain amplicons of the extracted DNA.

3. Results and discussion

The main objective is to investigate the decline and death of forest tree species in Namibia with a hope to develop management strategies to reduce the impact of the disease and help ensure the continued survival of an important component of the Namibian plant ecosystem for continued supply of ecosystem services. For this survey, for a number of specific fungal pathogens in key Namibian ecosystems. For this, the surveys conducted served as the groundwork for future research field of microbiology, plant pathology and plant protection. The results of this research helped to answer some key questions regarding fungal plant pathogens that occur in Namibia. Results (see Table 1) will expand our knowledge pertaining to the distribution, impact and origin of these pathogens. This knowledge is also essential to understand the working of these native ecosystems at a basic level, which is potentially important for the conservation of these ecosystems.

It is anticipated that the findings reported here will articulate into the Namibia's National Forest Programme aimed at promoting the sustainable use of the country's forests. In this national aim it is important to ensure forests that are healthy. If forest tree species are declining because of diseases, then sustainability becomes impossible. These data reported here assist in creating linkages with various sectors, such as the National Forestry Research Division in Okahandja in Namibia to promote forestry research, protect and conserve forest areas against destructive diseases and pests.

Most of the fungal species recorded in Table 1 have also been recorded elsewhere in the literature [9-15] as fungal pathogens affecting similar plant species that we investigated in this study or affecting other plant or crop species, which points to the fact that it is important to keep surveillance of plant disease as these fungal pathogens can easily be transported worldwide and be able to cause diseases or death in similar or other plant species.

Table 1. Fungi isolated from diseased indigenous plants in Namibia.

Area sampled	Tree species	Fungi isolated
Windhoek	**A. karroo**	*Acremonium* sp.
		Alternariacitri
		Botryosphearia sp.
		Curvulariapalescens
		Cytospora sp.
		Dreschlera sp.
		Fusarium spp.
		Microsphearopsis sp.
		Paecilomyceslilacinus
		Phoma spp.
		Trichodermakoningii
	A. hebeclada	A. citri
		Acremonium sp.
		Botryosphearia sp.
		Dreschlera sp.
		Fusarium spp.
		Microsphearopsis sp.
		Phoma spp.
	A. mellifera	*Fusarium* spp.
		Botryosphearia sp.
		Gliomastix sp.
		Paecilomycesvariotii
		Phomaglomerata.
	Acacia reficiens	*Gliocladium cibotii.*
		Alternaria tenuissima,
	Aloe zebrina	Ampelomyces spp
Dordabis	*A. mellifera*	*Botryosphaeria* sp.
		Coniothyrium sp.

		Cytospora sp
		Fusarium sp.
		Paecilomycesvariotii
		Phellinus sp.
		Phoma sp.
Grootfontein	*A. erioloba*	*Phellinus* sp.
	A. mellifera	*Botryosphaeria* sp.
		Fusarium sp.
		Phellinus sp.
		Phoma spp.
		Trichoderma sp.
Rundu	*A. erioloba*	*Phellinus* sp.
	A. mellifera	*Phellinus* sp.
Popa Falls	*A. mellifera*	*Phellinus* sp.
	S. guineense	*Chrysoporthe* sp.
Katima Mulilo	*A. erioloba*	*Phellinus* sp.
	S. guineense	*Chrysoporthe* sp.
Erongo	*W. mirabilis*	Botryosphaeria spp
Otjiwarongo	*Tylosema esculentum*	Phoma sorghina Alternaria tenuissima
		Alternaria alternata

Several fungal species have been isolated and some of them for the first time in diseased *Acacia* species and *Syzygium guineense* in Namibia. Visual inspections of disease symptoms on *A. karroo* trees in Windhoek showed that these trees had all been infested by a wood boring insect. This resulted in girdling of branches and subsequent wilt and death of these branches. Numerous fungal species were isolated from lesions ofinsect-infested trees (see Table 1). A few *A. hebeclada* trees were surveyed in Windhoek. Symptoms on these trees included stem cankers and branch die-back. No primary pathogens were, however, isolated from these trees. Those fungi isolated are opportunistic stress-associated pathogens capable of causing die-back and death of trees. Symptoms of foliage discoloration and apparent death of trees observed on *A. hebeclada* in the Rehoboth area were as the result of insect infestation. Caterpillars of an unidentified insect species, possibly of the Lepidoptera group were found feeding on the leaves and seedpods of these trees. Three sites with *A. mellifera* die-back were investigated. In Windhoek, tip defoliation and death of branch tips on *A. mellifera* trees caused by insects were observed. Extensive mortality of trees was observed on the farm near Dordabis, where numerous trees were sampled and isolated. On a farm in Grootfontein, branch die-back associated with a *Phellinus* sp. was the most common symptom observed, although many other fungi were also isolated (Table 1). *Chrysoporthe* sp. (previously *Cryphonectria*) was found on *S. guineense* trees around Popa Falls and Katima Mulilo. This fungus was found sporulating abundantly on stem and root cankers of trees.

4. Conclusions and recommendations

Several fungal species associated with diseases in *Acacia* and *Syzygium* trees in Namibia have been purified, stored and identified. There are ongoing surveys and confirmation studies of disease causation on other parts of the country. Population studies of the isolated fungi are underway. Possible pathogens and other fungi of interest are currently being identified to species level using morphology and DNA sequence data.

The results reported in this chapter can be used to develop a catalogue of fungi associated with crop and tree diseases in Namibia forest and possible control methods. In addition, the results can be used to devise long-term strategy for breeding for disease-tolerant forest species, especially those that have immediate commercial value, e.g., *Tylosema esculentum* and marula fruit plant. Brochures for disease alerts and awareness campaign on various trees diseases of Namibia can also be easily developed from these results. Copies of the brochures would then be made available for further dissemination to communities. The results here will help to ensure that the crops and native tree species are kept healthy. When the plants are healthy, food protection by crops and all the other ecosystem services and products are readily available to communities. This reduces the national burden on import of food and other plant-based services and products. The scientific data that have been generated will be used in plant health policy development.

References

1. Alberts AH. 2005. Threatening baobab disease in Nyae Nyae conservancy. Tsumkwe: Khaudum National Park.

2. Anonymous. 1991. Africa's favourite tree falls ill. *New Scientist*: 1784.

3. Chimwamurombe PM, Uzabakiriho J-D, Mapaure I. 2007. A survey of plant parasitic fungi occurring on selected native Namibian trees species of economic importance. *Biosciences, Biotechnology Research Asia* 4(1): 23-28.

4. De Villiers PC. 1951. Die Kremetartboom (*Adansonia digitata*, Linn). *Journal of South African Forestry* 20: 9-18.

5. Piearce GD, Calvert GM, Sharp C, Shaw P. 1994. Sooty Baobabs - Disease of drought? In Research Paper No. 6. Harare: Forest Research Centre.

6. Roux J. 2002. Baobab mortality in the Musina nature reserve – 14-16 March 2002 a pilot study.

7. Slippers B, Smit WA, Crous PW, Coutinho TA, Wingfield BD, Wingfield MJ. 2007. Taxonomy, phylogeny and identification of Botryosphaeriaceae associated with pome and stone fruit trees in South Africa and other regions of the world. *Plant Pathology* 56: 128-139.

8. Smith H, Wingfield MJ, Petrini O. 1996. *Botryosphaeria dothidea* endophytic in *Eucalyp- tus grandis* and *Eucalyptus nitens* in South Africa. *Forest Ecology and Management* 89: 189-195.

9. Kanyomeka L, Nalisa M, Mapaure I, Chimwamurombe PM. 2007. First report of *Alternaria compacta* occurring on Namibian *Acacia karoo*. *Journal of Pure and Applied Mi- crobiology* 1(2): 177-180.

10. Chimwamurombe PM, Mapaure I, Haufiku MS. 2009. DNA-based fungal identification: First report of double infection of *Alternaria tenuissima* and *Ampelomyces* species on malformed pods of *Acacia karoo* in Namibia. *Asian Journal of Microbiology, Biotech- nology and Environmental Sciences* 11(3): 665-669.
11. Nashidengo E, Uzabakiriho JD, Chimwamurombe PM. 2011. Fungi associated wall bizarre gall formations on *Acacia reficiens* in the Numas river area in Namibia. *Journal of Pure and Applied Microbiology* 5(2): 617-621.
12. Garas SL, Uzabakiriho JD, Chimwamurombe PM. 2012. Isolation and identification of fungal species associated with gall formation on *Acacia mellifera* in the western Windhoek. *Journal of Pure and Applied Microbiology* 6 (2): 713-716.
13. Kido LR, Uzabakiriho JD, Chimwamurombe PM. 2012. Isolation and identification of pathogenic fungi associated with *Aloe zebrina* flower malformation. *Journal of Pure and Applied Microbiology* 6(1): 125-129.
14. Martin P, Uzabakiriho JD, Chimwamurombe PM. 2012. Molecular identification of leaf endophytic fungi affecting marama bean leaves in Otjiwarongo area of Namibia. *Journal of Pure and Applied Microbiology* 6(2): 665-670.
15. Takundwa MM, Ruzvidzo O, Uzabakiriho, JD, Titus PE, Chimwamurombe PM. Molecular identification of fungal species associated with leaf lesions of Marama bean seedlings in Namibia. *African Journal of Microbial Research* 9(3): 194-200.

Index